Electrical Vehicle Powertrains: Designs and New Paradigms

Electrical Vehicle Powertrains: Designs and New Paradigms

Edited by
Lucy Dickinson

| STATES |
ACADEMIC PRESS
www.statesacademicpress.com

Published by States Academic Press,
109 South 5th Street,
Brooklyn, NY 11249, USA

ISBN: 978-1-63989-173-3

Cataloging-in-Publication Data

Electrical vehicle powertrains : designs and new paradigms / edited by Lucy Dickinson.
    p. cm.
Includes bibliographical references and index.
ISBN 978-1-63989-173-3
1. Electric vehicles--Power trains. 2. Automobiles--Power trains.
3. Electric vehicles--Power trains--Design and construction.
4. Motor vehicles. I. Dickinson, Lucy.
TL272.5 .E44 2022
629.254 9--dc23

For information on all States Academic Press publications
visit our website at www.statesacademicpress.com

STATES
ACADEMIC PRESS

Contents

Permissions

List of Contributors

Index

Preface

The purpose of the book is to provide a glimpse into the dynamics and to present opinions and studies of some of the scientists engaged in the development of new ideas in the field from very different standpoints. This book will prove useful to students and researchers owing to its high content quality.

The motor vehicle which comprises a source of propulsion such as electric motor, and a drivetrain system that is used to transfer this energy into the forward movement of vehicles, is known as an electric powertrain. Environmental aspects, socio-economic factors, resilience, energy efficiency, cost of recharging, stabilization of the grid, and heating of electric vehicles are some of the advantages of electrical vehicle power trains. They have very good power-to-weight-ratios as they do not carry heavy internal combustion engines. The various studies that are constantly contributing towards advancing technologies and evolution of this field are examined in detail in this book. It presents the complex subject of electrical vehicle powertrains in the most comprehensible and easy to understand language. This book is an essential guide for both academicians and those who wish to pursue this discipline further.

At the end, I would like to appreciate all the efforts made by the authors in completing their chapters professionally. I express my deepest gratitude to all of them for contributing to this book by sharing their valuable works. A special thanks to my family and friends for their constant support in this journey.

Editor

Estimation Techniques for State of Charge in Battery Management Systems on Board of Hybrid Electric Vehicles Implemented in a Real-Time MATLAB/SIMULINK Environment

Roxana-Elena Tudoroiu, Mohammed Zaheeruddin,
Sorin-Mihai Radu and Nicolae Tudoroiu

Abstract

The battery state-of-charge estimation is essential in automotive industry for a successful marketing of both electric and hybrid electric vehicles. Furthermore, the state-of-charge of a battery is a critical condition parameter for battery management system. In this research work we share from the experience accumulated in control systems applications field some preliminary results, especially in modeling and state estimation techniques, very useful for state-of-charge estimation of the rechargeable batteries with different chemistries. We investigate the design and the effectiveness of three nonlinear state-of-charge estimators implemented in a real-time MATLAB environment for a particular Li-Ion battery, such as an Unscented Kalman Filter, Particle filter, and a nonlinear observer. Finally, the target to be accomplished is to find the most suitable estimator in terms of performance accuracy and robustness.

Keywords: Li-Ion battery state-of-charge, state estimation, unscented Kalman filter estimator, particle filter estimator, nonlinear observer estimator, battery management system

1. Introduction

We are currently seeing a significant increase in global environmental pollution, with immediate repercussions on air, water and soil quality. More precisely, especially in the developed countries around the world the environmental pollution has reached scary limits. Related to this it is worth to mention the presence of a significant amount of hydrocarbons pollutants (benzene, toluene and xylene) in the emissions of the vehicles equipped with gasoline or diesel engines that are characterized by a variable toxicity depending on the chemical composition of the exhaust gases. Furthermore, these toxic substances are propagated through the air from one region of the world to another one and surrounds countries and continents becoming a global phenomenon, consisting of irreversible pollution of water, air and soil at the planetary scale.

Therefore, the need to conceive and implement new environmental conservation strategies at the global scale is required. Also, a changing in the thinking of the people about a significant reduction in energy consumption without sacrificing the comfort is crucial. In these circumstances there is a real hope that with the current technology available could stop the global destruction of the environment. Moreover, the new strategies based on electrical energy consumption assure a sustainable development of each community are critical to achieve a clean and efficient urban or rural transportation. As a viable solution to the global energy shortage and growing environmental pollution is the use of the electric vehicles (EVs) [1]. Nowadays, the electric vehicles (EVs) including hybrid electric vehicles (HEVs), plug-in hybrid electric vehicles (PHEVs), and pure battery electric vehicles (BEVs) are gaining popularity in automotive industry and will dominate soon the clean vehicle market [2]. Related, in [2] is mentioned also that by 2020, it is expected that *"more than half of new vehicle sales will likely be EV models"*, with the batteries playing *"the key and the enabling technology to this revolutionary change"*. They are conceived *"to handle high power (up to a hundred kW) and high energy capacity (up to tens of kWh) within a limited space and weight and at an affordable price"* [2]. The most advanced and promising battery technologies existing in EVs manufacturing automotive industry are the nickel-metal hydride (NiMH), lithium-ion (Li-Ion) and nickel-cadmium (NiCad) batteries considered as the most suitable for HEVs/PHEVs/EVs all over the world.

They have a great potential to reduce greenhouse and other exhaust gas emissions, and require extensive research efforts and huge investments [2].

Nevertheless, amongst them the most promising power source with a great potential to be developed and to get a wide application in the future on the EVs market is the Li-Ion battery recommended by its light weight, high energy density, tiny memory effect, and relatively low self-discharge compared to its strong competitors, namely Ni-Cad and Ni-MH batteries, as is mentioned in [1, 2].

Additionally, the newest Li-Ion batteries are safer and less toxic than the same batteries in competition. Due to the diversity and the complexity of EVs field we limit our case study only to HEVs applications since we have got some research experience in modeling, control and the estimation strategies related to this field.

Therefore, one of the main objectives of this research work is to disseminate the most relevant results obtained until now in this area and to share some interesting ideas with our readers. The Li-Ion battery is a main component integrated in the battery management system (BMS) of a HEV that is responsible for *"improving the battery performance, prolonging battery life, and ensuring its safety"*, as is mentioned in [1, 2]. This desideratum is achieved by the BMS through recording continuously the main parameters of the Li-Ion battery and performing an accurate estimation of its state-of-charge (SOC). An accurate SOC estimation is a vital operation to be performed by the BMS of the HEV in order to prevent the dangerous situations when the battery is over-charged or over-discharged, and to improve considerably the battery performance [1]. More precisely, the battery SOC is an inner state of a battery that can be defined as the available capacity of a battery, as a percentage of its rated capacity [1–3]. Its estimation is an essential operational condition parameter for battery management system (BMS) but it cannot be measured directly [1]. The estimation of Li-Ion battery SOC value is based on the measurable data set of the battery parameters, mainly the current, voltage, and temperature by using several estimation strategies, implemented in real-time MATLAB/SIMULINK platform that includes many real-time features [3–11]. All SOC estimation strategies are model-based, and can be grouped in Kalman Filter (standard, extended, unscented, particles filter), as those developed and implemented in real-time in [3–11], linear, nonlinear and sliding mode observers estimators, including also fuzzy improved versions, well documented in [1, 8–10]. The environmental impact is a key issue on the enhancing the battery technologies, as is mentioned in [12].

Definitely, the selection criteria of the specific chemistry battery to be integrated in BMS structure of a HEV are the cost, the specific power and energy, cycle life, and the presence of poisonous heavy metals [2, 12]. A complete literature review on life cycle assessment (LCA) of HEVs since 1998 until 2013 has been conducted in [12]. In our research we are only focused on the technical aspects, such as battery modeling and developing the most suitable estimation techniques of battery SOC.

The remainder of this chapter is organized as follows. In Section 2, the widely-used 2RC-series cells Li-Ion battery equivalent model circuit (EMC) is introduced and the state space equations are derived. In Section 3, is proposed for design and implementation in real-time three nonlinear estimators, namely an Unscented Kalman Filter (UKF), Particle Filter (PF), and a nonlinear observer estimator (NOE).

The simulation results and the performance analysis of the proposed estimators are presented in Section 4. Section 5 concludes the book chapter.

2. The continuous and the discrete time state-space Li-Ion battery model representation

In this section we introduce a generic model capable to describe accurately the dynamics of Li-Ion battery, based on the same set of first-order differential equations in a state-space representation. For simulation purpose, a specific Li-Ion battery model is considered to prove the effectiveness of the proposed SOC estimation strategies. This model can be obtained from the generic model by changing only the values of the model parameters in state-space equations. For our case study we choose the widely-used 2RC - series cells Li-Ion battery equivalent model circuit (EMC) as model-based support, well documented in [1, 8, 11].

2.1. Li-Ion battery terminology

In this subsection we introduce briefly the same terminology as in [2] to introduce the most used terms in this chapter that characterize the Li-Ion battery architecture and its performance.

2.1.1. Li-Ion battery characterization architecture terminology

A single battery cell is a complete battery with two current leads and separate compartment holding electrodes (positive (+) and negative (−)), separator, and electrolyte. A few cells are connected in series, parallel or in series - parallel combinations, either by physical attachment or by welding in between cells to form a battery module. Similar, several modules are connected to form a battery pack placed in a single compartment for thermal management. A HEV may have more than one batteries packs placed in a different location of the car.

2.1.2. Li-Ion battery performance terminology

A battery cell is fully charged when its terminal voltage reaches the maximum voltage limit value after being charged at infinitesimal current levels, for example a maximum limit voltage value of 4.2 V at room temperature (25°C). A battery terminal voltage value greater than this limit corresponds to a dangerous over-charging operating condition. Similar, a battery cell is fully discharged when its terminal voltage reaches the minimum voltage limit value after being drained at infinitesimal current levels, for example a minimum limit voltage value of 3.0 V at room temperature. A battery terminal voltage value smaller than this limit corresponds to a dangerous over-discharging operating condition. The capacity of a battery is measured in Ampere-hours (Ah) defined as the total charge that can be discharged from a fully charged battery under specified conditions. The rated Ah capacity is the nominal capacity of a fully charged new battery under the conditions predefined by the catalog specifications of the battery, e.g. the nominal condition could be defined as room temperature 25°C and battery discharging is at 1/25 C-rate. The C-rate is used to represent a charge or discharge rate equal to the capacity of a battery in 1 hour, e.g. for a 6 Ah battery, 1C-rate is equal to charge or discharge the battery at a constant current value of 6 A; in the same way, 0.1C-rate is equivalent to 0.6 A, and 2C-rate for charging or discharging the battery at a constant current value of 12 A. The power density of the battery is an important criterion of battery selection that is defined as the peak power per unit volume of a battery (W/l). In the battery model dynamics, the effect of its internal resistance is significant. It is defined as the overall equivalent resistance within the battery. Also, it is worth to mention that its value varies during the charging and discharging battery cycles, and moreover may vary as the operating condition changes. The peak power according to the U.S. Advanced Battery Consortium (USABC)'s definition is given by:

$$P = \frac{2V_{OC}^2}{9R} \tag{1}$$

where V_{OC} is the open-circuit voltage (OCV) and R is the battery internal resistance.

The peak power is defined at the condition when the terminal voltage is 2/3 of its OCV. The SOC of the battery provides an important feedback about the state of health (SOH) of the battery and its safe operation. SOC is defined as battery available capacity expressed as a percentage of its rated capacity. More precisely, the SOC can be defined as the remaining capacity of a battery and it is affected by its operating conditions such as load current and temperature:

$$SOC = \frac{Remaining\ capacity}{Rated\ capacity} \qquad (2)$$

The SOC for a fully charged battery is 100% and for an empty battery is 0%, defined for a discharging cycle, when discharging battery current is positive, as:

$$SOC(t) = 100\left(1 - \frac{\eta}{Ahnom}\int_0^t i(\tau)d\tau\right) \ (\%), \ \ i(\tau) \geq 0 \qquad (3)$$

where η is the coulombic efficiency of the charging or discharging battery cycle, $Ahnom$ represents the nominal battery capacity, $i(\tau)$ is the instantaneous value of charging ($i(t) \leq 0$) or discharging ($i(t) \geq 0$) battery current.

The relation (3) can be also written as a first order differential equation that further it will be used for SOC state estimation:

$$\frac{d}{dt}(SOC(t)) = -100\frac{\eta \times i(t)}{Ahnom}, i(t) \geq 0 \qquad (4)$$

The SOC is a critical condition parameter for battery management system (BMS), often affected by its operating conditions such as load current and temperature; consequently, an accurate estimation of SOC is very important, since it is the key issue for the healthy and safe operation of batteries. The depth of discharge (DOD) is used to indicate the percentage of the total battery capacity that has been discharged at time t, defined as:

$$DOD(t) = 100(1 - SOC(t)) \ (\%) \qquad (5)$$

The SOH of the battery is defined as the ratio of the maximum charge capacity of an aged battery to the maximum charge capacity when this battery was new. Its life cycle (number of cycles) is given by the number of discharging-charging cycles that the battery can withstand at a specific DOD (normally 80%) before it fails to meet the desired performance criteria. The actual operating life of the battery is affected by the charging and discharging rates, DOD, and by the temperature. The higher the DOD is the shorter will be the cycle life. To attain a higher life cycle, a larger battery is required to be used for a lower DOD during normal operating conditions.

The Battery Management System is an integrated battery structure consisting of measurement sensors, controllers, serial communication, and computation hardware with software algorithms such as Proportion Integral Derivative (PID) and adaptive control laws, Kalman filters estimators, adaptive or sliding mode observers designed to decide the maximum charging/ discharging cycles current and the duration from the estimation of SOC and SOH of the battery pack, as is shown in [11, 13–15].

2.1.3. Battery management systems: architecture, trends, functions, monitoring, faults detection, hardware and software components

A BMS as an important connector between the battery and the HEV plays a vital role in improving battery performance and optimizing vehicle operation in a safe and reliable manner, as is mentioned in [13]. Nowadays, the trend of the BMSs is a rapid growth of the EV and HEV market, thus it is essential for the automotive industry to develop a comprehensive and mature BMSs. As is stated in [13]"*the U.S. Council for Automotive Research (USCAR) and the U.S. Advanced Battery Consortium (USABC) have set minimum goals for battery characteristics for the long-term commercialization of advanced batteries in EVs and hybrid electric vehicles (HEVs)*". Furthermore, to increase the market segment of EVs and HEVs automotive industry, the main concerns such as reliability and safety remain constantly for both of them, battery technology (BT) and BMS.

The BMS hardware and software components, the safety circuitry incorporated within the battery packs play an important role to monitor, control, compute and to show continually the safety state, the SOC, SOH, as well as the longevity of the battery.

Moreover, in [13] is emphasized one of the most dangerous situations such as the ignition of a Li-Ion battery during overcharging operating conditions, due to the volatility, flammability and entropy changes. Also, after repeated over-discharging cycles, the battery cell capacity is reduced significantly, due to irreversible chemical reactions, and thus the BMS needs to monitor and control constantly the Li-Ion battery state. Whenever any abnormal conditions happen, such as self-discharge leakage current through the insulation resistance of the battery, well-known as ground insulation resistances of the negative and positive bus of BMS R_p, and *Rn* respectively, over-voltage or overheating operating conditions are identified, and in a very short time the BMS should notify the user and to execute the preset correction procedures [13, 16]. In addition to these foremost functions, the BMS also monitors the system temperature to provide a better power consumption scheme, and communicates with individual components and operators [13]. Technically, a comprehensive BMS is equipped with the most suitable hardware and software components of the newest generation [14, 15], integrated in the HEVs and EVs structure to accomplish the main following functions: (1) real-time monitoring of battery states by a performing data acquisition system of external signals (i.e. voltage, current, cell temperature etc.); (2) ensure user safety protection, and extend the battery life; (3) using performing and intelligent algorithms (as for example, genetic, fuzzy logic, neural networks and expert systems based on artificial intelligence) has the ability to estimate and monitor the battery internal parameters and states (i.e. the DC resistance, insulation resistances R_p, and *Rn*, polarization voltage, maximum available capacity, SOC, SOH, etc.); (4) the ability to prevent over-charge or over-discharge of the battery; (5)efficient battery energy utilization, thermal management and SOC cell balancing; (6) delivery of battery status and authentication to a user interface; (7) the ability to communicate with vehicle controller and all other components [13, 16]. In order to achieve these objectives, researchers focus on battery modeling, SOC estimation, consistency evaluation and equalization, such as stated in [16].

Furthermore, to increase the effectiveness of BMS in [13] is proposed a new BMS with the following categories of components:

A. Hardware with the following components:

Safety circuitry; Sensor system; Data acquisition; Charge and discharge control; Communication; Thermal management.

The sensor system block integrates different sensors capable to monitor and measure battery parameters including cell voltage, battery temperature, and battery current. The new proposed BMS in [13] it seems to have a lot of improvements in terms of current safety circuitry designs that can be easily implemented, amongst them is worth to mention the addition of accurate alarms and controls to prevent overcharge, over-discharge, and overheating. Therefore, the external signals such as the current, voltage, and temperature must be measured to improve the capability of state tracking in real life applications [13]. Data acquisition (DAQ) block in conjunction with the data storage block are critical parts for the software in the BMS to analyze and build a database for system modeling and estimation algorithms development.

The charge and discharge control block is integrated in hardware architecture of the BMS to implement the charge-discharge protocol. A variable resistor may be necessary to help balance cells or perform internal resistance measurements. In [13] is mentioned also that the "cell balancing control is still a critical design feature with room for improvement in order to equalize the battery pack and estimate the battery status in an efficient way". Most subsystems in a BMS are stand-alone modules, and hence, data transfer throughout the BMS is required. Communication through a CAN Bus is a major way to transfer data between the stand-alone subsystems within the BMS [13].

Furthermore, the recent development of smart batteries within the context of innovative electronics and artificial intelligence, more data can be collected to communicate with the user and the charger through the microchips incorporated within the battery [13]. In addition, "wireless and telecommunication techniques are gradually being incorporated into charging systems that facilitate communication between the battery and the charger". Finally, the last thermal management hardware module proved during the time that its integration in a BMS is critical to be integrated since the temperature variations between the cells have a great impact on cell imbalance, reliability and performance [13].

B. Software with the following components:

SOC estimation and monitoring; SOH estimation and monitoring; Cell balancing; Fault Detection and Isolation; User interface.

The software compartment integrated in BMS structure is in reality such as an artificial "brain" that controls all hardware operations and examines all sensors data for making decisions and to implement in real-time linear, nonlinear and intelligent SOC, SOH estimators, and also FDI techniques. As is mention in [13], "the switch control, sample rate monitoring in the sensor system block, cell balancing control, and even dynamic safety circuit design should be handled by the software of a BMS". In addition, the software components assure online data processing, and also reliable and robust automated data analysis by controlling battery functions, representing an essential factor to perform successfully the SOC, SOH estimation and fault detection and isolation algorithms. The robust SOC and SOH estimators are design to integrate also a capability assessment of the life status of the battery and to *"sets the operating limits according to state-of-the-art algorithms, such as fuzzy logic, neural networks, state-space-based models, and so on "*[13]. Most of the soft battery faults are detected through online data processing by means of an intelligent data analysis that provides information about the occurrence of the faults and their persistence by fault warnings, and also indicates abnormal limits of operating conditions. The history of the past measurements data set is recorded and provides the pre-alarm condition before the possible battery faults. In [17, 18] is proposing a real-time diagnosis approach that detects, isolates, and estimate specific sensor faults, such as battery voltage, current temperature, and fan motor. The sensor fault estimation can provide great benefits for enhancing the reliability of BMSs. Also, the sensor fault estimation can provide fault-tolerance capability to the BMS by allowing it to continue in degraded and in safe mode even after the sensor faults occur, as is mentioned in [17]. Failure in Li-Ion batteries can be also attributed to a combination of manufacturing defects, safety component failure, or human operating errors, as is specified in [18]. Furthermore, in [18] are developed some methods for realistic fault injection. Two types of sensor faults are considered in [18], namely an intermittent signal loss due to faulty wiring connections and sensor bias resulting from time or temperature drift. For the fan motor, the only fault considered is a total motor failure, and thus no longer provide cooling to the battery. The occurrence of the realistic battery faults are established based on the preset thresholds. These thresholds are computed as a minimum values that assure the prevention of the false alarms, based on the minimization of the residual error probability when occurs a fault, by studying the probability density functions (PDF)) of the healthy and faulty signals, such is shown in [18]. The PDfs are obtained through extensive experimental data collection or simulations on the system.

Roughly, for many faults scenarios to compute the magnitude of the thresholds we can use also the statistics of the residual errors for healthy system between the measured values data set and its estimated values, such as the mean (μ_X) and the standard deviation (σ) calculated in steady state. The threshold can be calculated by addition to the mean (μ_X) of residual error approximatively two times the standard deviation (2σ). The decision for occurrence of the fault is given by comparison of the residual error and the computed threshold. The tuning of the threshold to avoid the false alarms can be done by well-known trials and error method. The user interface block must display the essential information of the BMS to the users.

2.2. Li-Ion battery EMC description

The equivalent model circuits (EMCs) are simple electric circuits consisting of voltage sources, resistors and capacitors networks, commonly-used to simulate the dynamic behavior of the Li-Ion batteries [1, 8, 11]. The role of the resistor-capacitor (RC) series networks integrated in EMCs is to improve models' accuracy, and also to increase the structural complexity of the models [1, 8]. In this research paper we chose for Li-Ion battery model design and implementation stages an equivalent second order model RC circuit. It consists of a resistor R, two series RC cells (R_1C_1, R_2C_2), and an open-circuit voltage (OCV) source dependent of SOC, i.e. OCV (SOC), as is shown in **Figure 1(a)**, in NI Multisim 14.1 editor. In EMC architecture, R denotes the internal ohmic resistance of the Li-Ion battery, the parallel cells (R_1, C_1), (R_2, C_2) connected in series denote the electrochemical polarization resistances and capacitances, for which $T_1 = R_1C_1$, $T_2 = R_2C_2$ can imitate the fast polarization time constant, and the slow polarization time constant for the voltage recovery of the Li-Ion battery respectively {14}. Also, $i(t)$ is the battery instantaneous value of the direct current (DC) flowing through the OCV source, and $v(t)$ represents the measured battery instantaneous value DC voltage. The main benefit of the RC second order EMC architecture selection is its simplicity and the ability to be implemented in real-time applications with acceptable range of performance [1, 8].

Also "*this choice is due to the early popularity of BMS for portable electronics, where the approximation of the battery model with the proposed EMC is appropriate*", as is mentioned in [1, 8]. Related, this approach has been extended easily to Li-Ion batteries for automotive industry and for many other similar energy storage applications [11]. Since the EMC is used in following sections to design and implement in real-time all three proposed nonlinear state estimators, now it is essential to precise the main factors that affects the battery parameters, especially the internal and insulation resistances and their impact on the battery dynamics in a realistic operating conditions environment. The internal resistance of the battery is affected by the following factors: conductor resistance, electrolyte resistance, ionic mobility, separator efficiency, reactive rates at the electrodes, and concentration polarization, temperature effects and changes in SOC. When a battery fails, it is typically since it has built up enough internal resistance that it can no longer supply a useful amount of power to an external load, according to the maximum power transfer between the source and the load. The insulation positive and negative resistances of the battery are denoted by R_n, and R_p respectively, and are shown in the diagram from **Figure 1(b)**. The main objective is to get a good insight on the impact of the insulation

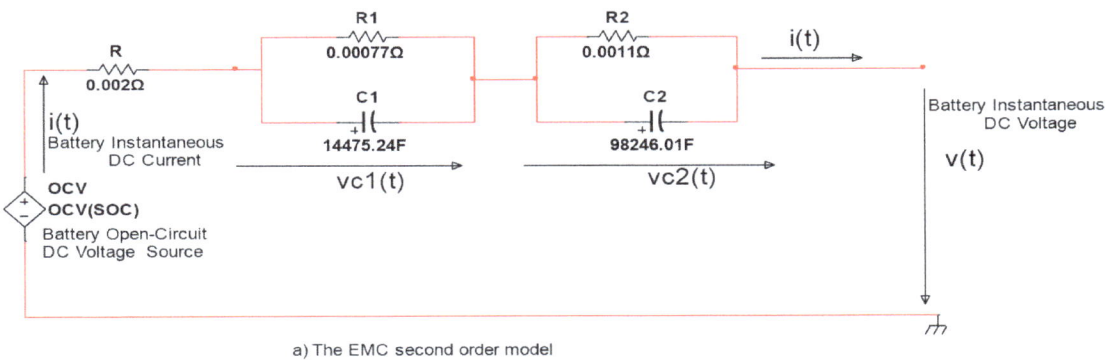

a) The EMC second order model

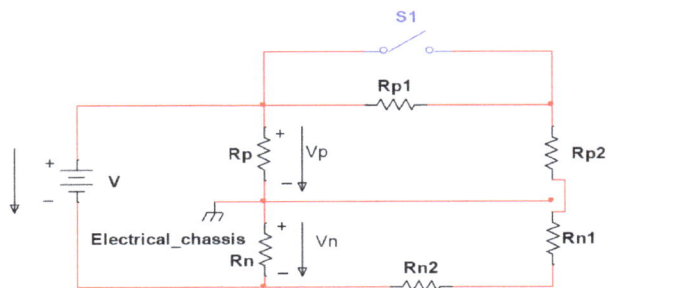

b) Diagram of the passive ground detection method of insulation resistances Rn and Rp

Figure 1. The second order RC equivalent model circuit (EMC) 1(a), and the insulation resistances Rn and Rp 1(b) of Li-Ion battery, represented in NI Multisim 14.1 editor (see for details [16]).

resistances on the battery dynamics, as is described in detail in [16]. Also, in [16] the HEV is considered as a *"complex production of mechanical-electrical integration"*, for which the power supply typically being in the range 100–500 V is obtained by means of several series battery packs. Amongst the BMS hardware devices consisting of high voltage components, the traction battery, electrical motor, energy recycle device, the battery charger and its auxiliary device deal with a large current and insulation [16], thus insulation issue must be under consideration since the design stage. As is stated in [16], the poor working conditions, such as shaking, corrosion, changes in temperature and humidity, *"could cause fast aging of the power cable and insulation materials, or even brake the insulation, which would decrease the insulation strength and endanger personnel"*. Thus, needs to ensure safety operating conditions for personnel are required to evaluate the insulation conditions for entire HEV's BMS.

The National Standard (NS) 18384.3-2001, stipulates several safety requirements for HEVs, especially for insulation resistance state, measurement method [16]. According to NS, the insulation state of an EV is evaluated according to the ground insulation resistance of the DC positive and negative bus, V_p, and V_n respectively, as is shown in **Figure 1(b)** [16]. The definition of traction battery insulation resistance in NS *"is the relative resistance to maximum leakage current (in the worst condition) where there is a short between the traction battery and ground (electric chassis)"* [16]. Also, *"under the conditions that the maximum AC voltage is less than 660 V, the maximum DC voltage is less than 1000 V and the car weight is less than 3500 Kg, the requirements of the high voltage security are as follows: (1) personnel's security voltage is less than 35 V, or the product of the contact current with a person and the duration of time is less than 30 mA s; insulation resistance divided by the battery rated voltage should be more than 100 ΩV^{-1}, and preferably more 500 ΩV^{-1}"*, as is formulated in [16]. Thus, to ensure the insulation security of on-board BMS, it is necessary to detect the insulation resistance and raise an alarm in time. Currently, the insulation measurement methods, such as passive ground detection and active ground detection, include the AC voltage insulation measurement method, and the DC voltage insulation measurement method [16]. The passive ground detection insulation measurement method principle is presented in **Figure 1(b)** where the presence of leakage current is detected by discovering the resultant magnetic field generated by AC voltage around the mutual inductor.

2.3. Li-Ion battery continuous time state-space representation

By simple manipulations of Ohm's, current and voltage Kirchhoff's laws applied to the proposed Li-Ion battery's EMC can be written the following first order differential equations that are most suitable to capture the entire dynamics of EMC in time domain:

$$\frac{dv_{C_1}}{dt} = -\frac{1}{R_1 C_1} v_{C_1}(t) + \frac{i(t)}{C_1}$$
$$\frac{dv_{C_2}}{dt} = -\frac{1}{R_2 C_2} v_{C_2}(t) + \frac{i(t)}{C_2}$$
$$\frac{d(SOC)}{dt} = -\frac{\eta}{Q_{nom}} i(t)$$
$$v(t) = OCV(SOC(t)) - v_{C_1}(t) - v_{C_2}(t) - Ri(t)$$

(6)

By defining the following state variables $x_1 = v_{C_1}(t)$, $x_2 = v_{C_2}(t)$, and $x_3 = SOC(t)$, representing the polarization voltages of the two RC cells, and considering as battery input u the DC instantaneous current that flows through battery, $u = i(t)$, and as a battery output y, designating the terminal battery DC instantaneous voltage, i.e. $y = v(t)$, the Eq. (6) can be written in a state-space representation, as follows.

$$\frac{dx_1}{dt} = -\frac{1}{T_1}x_1 + \frac{1}{C_1}u$$
$$\frac{dx_2}{dt} = -\frac{1}{T_2}x_2 + \frac{1}{C_2}u \qquad (7)$$
$$\frac{dx_3}{dt} = -\frac{\eta}{Q_{nom}}u$$
$$y = OCV(x_3) - x_1 - x_2 - Ru$$

where $T_1 = R_1 C_1 [s]$, $T_2 = R_2 C_2 [s]$ represent the time constants of the both polarization RC cells, and the OCV combines three additional well-known models, as defined and used in [4, 8], given by:

$$OCV(x_3(t)) = h(x_3(t)) = K_0 - K_1 \frac{1}{x_3(t)} - K_2 x_3(t) + K_3 \ln(x_3(t)) + K_4 \ln(|1 - x_3(t)|) \qquad (8)$$

For simulation purpose, in order to validate the proposed battery model, as well as to prove the effectiveness of the SOC estimation strategies developed in Section 3, we consider as the most suitable nominal values for Li-Ion battery EMC parameters and OCV coefficients the same values that are carefully chosen for model validation in [8] as follows

1. Li-Ion battery ECM parameters:
   - The battery internal ohmic resistance (slightly different for charging and discharging cycles), R = 0.0022 [Ω] (ohms), the first RC cell polarization resistance and capacitance, R_1 = 0.00077 [Ω], C_1 = 14475.24 [F], and the second RC cell polarization resistance and capacitance, R_2 = 0.0011 [Ω], C_2 = 98246.01 [F] (Farads).

2. Li-Ion battery characteristics:
   - The value of the battery capacity, Q = 6 Ah (Amperes hours), the voltage nominal value of the battery, V_{nom} = 3.6 V (Volts), and the coulombic efficiency, η = 1, for charging cycle, and η = 0.85, for discharging cycle

3. The OCV coefficients:
   - K_0 = 4.23, K_1 = 3.86E-05, K_2 = 0.24, K_3 = 0.22, K_4 = −0.04

The $OCV(x_3(t))$ is a nonlinear function of SOC, i.e.x_3, similar as in [1, 3, 4, 8–11], increasing the accuracy of the Li-Ion battery EMC model, known as the EMC combined model, one of the most accurate formulation by combining Shepherd, Unnewehr universal, and Nernst models introduced in [4, 8]. The tuning values of model parameters $(K_0, K_1, K_2, K_3, K_4)$ are chosen to fit the model to the manufacture's data by using a least squares curve fitting identification method OCV = h (SOC), as is shown in [3, 4, 8], where the OCV curve is assumed to be the average of the charge and discharge curves taken at low currents rates from fully charged to fully discharged battery [3, 4, 6, 8]. Also, by using low charging and discharging DC currents can be minimized the battery cell dynamics. A simple offline (batch) processing method for parameters calculation is carried out in [3, 4, 6, 8].

2.4. Li-Ion battery discrete time state-space representation

We introduce now the following new notations:

$x_1(k) = x_1(kT_s), x_2(k) = x_2(kT_s), x_3(k) = x_3(kT_s)$, for state variables, $u(k) = u(kT_s)$, $y(k) = y(kT_s)$ for input current profile, and output battery terminal DC samples voltage respectively, commonly used for discrete time description with the sampling time $T_s [s]$, similar as in [1, 3, 4, 6, 8–11].

With these notations, a discrete time state space representation of the combined EMC Li-Ion battery model is obtained::

$$
\begin{bmatrix} x_1(k+1) \\ x_2(k+1) \\ x_3(k+1) \end{bmatrix} = \begin{bmatrix} 1-\dfrac{T_s}{T_1} & 0 & 0 \\ 0 & 1-\dfrac{T_s}{T_2} & 0 \\ 0 & 0 & 1 \end{bmatrix} \begin{bmatrix} x_1(k) \\ x_2(k) \\ x_3(k) \end{bmatrix} + \begin{bmatrix} \dfrac{T_s}{C_1} \\ \dfrac{T_s}{T_2} \\ -\dfrac{\eta T_s}{Q_{nom}} \end{bmatrix} u(k) \tag{9}
$$

$$
y(k) = h(x_3(k)) - x_1(k) - x_2(k) - Ru(k) = OCV(SOC(k)) - x_1(k) - x_2(k) - Ru(k) \tag{10}
$$

Further, a compact discrete time of combined ECM Li-Ion battery state space representation (9) can be written in the following matrix form:

$$
x(k-1) = Ax(k) + Bu(k)
$$
$$
y(k) = Cx(k) + Du(k) + \Psi(x_3(k))
$$

$$
x(k) = [x_1(k)\ x_2(k)\ x_3(k)]^T, A = \begin{bmatrix} 1-\dfrac{T_s}{T_1} & 0 & 0 \\ 0 & 1-\dfrac{T_s}{T_2} & 0 \\ 0 & 0 & 1 \end{bmatrix}, B = \begin{bmatrix} \dfrac{T_s}{C_1} \\ \dfrac{T_s}{T_2} \\ -\dfrac{\eta T_s}{Q_{nom}} \end{bmatrix}, C = [-1\ -1\ -K_2],
$$

$$
\tag{11}
$$

$$
D = -R, \Psi(x_3(k)) = K_0 - K_1 \frac{1}{x_3(k)} + K_3 \ln(x_3(k)) + K_4 \ln(|1 - x_3(k)|) \tag{12}
$$

where the nonlinear function $\Psi(x_3(k))$ can be further linearized around an operating point to get a linear Li-Ion battery combined EMC, easily to be implemented in real-time. To analyze the behavior of the proposed Li-Ion battery EMC for different driving conditions such as urban, suburban and highway, some different current profiles tests will be introduced in the next subsection.

2.5. Li-Ion battery equivalent model in ADVISORY MATLAB platform—case study

For EMC validation purpose we compare the results of the tests using a NREL with two capacitors already integrated in an Advanced Vehicle Simulator (ADVISOR) MATLAB platform, developed by US National Renewable Energy Laboratory (NREL), as is shown in [9, 10]. The NREL Li-Ion battery model approximates with high accuracy the Li-Ion battery model 6 Ah and nominal voltage of 3.6 V, manufactured by the company SAFT America, as is mentioned in [9, 10]. Also, for simulation purpose and comparison of the tests results, the EMC battery model is incorporated in a BMS' HEV, and its performance is compared to those obtained by a particular Japanese Toyota Prius, selected as an input vehicle in ADVISOR MATLAB platform, under standard initial conditions and the setup shown in **Figure 2**.

Among different driving speed cycles for a large collection of cars provided by the ADVISOR US Environmental Protection Agency (EPA), in our case study for Toyota Prius HEV car, the speed profile is selected as an Urban Dynamometer Driving Schedule (UDDS), as is shown in **Figures 3** and **4**, respectively.

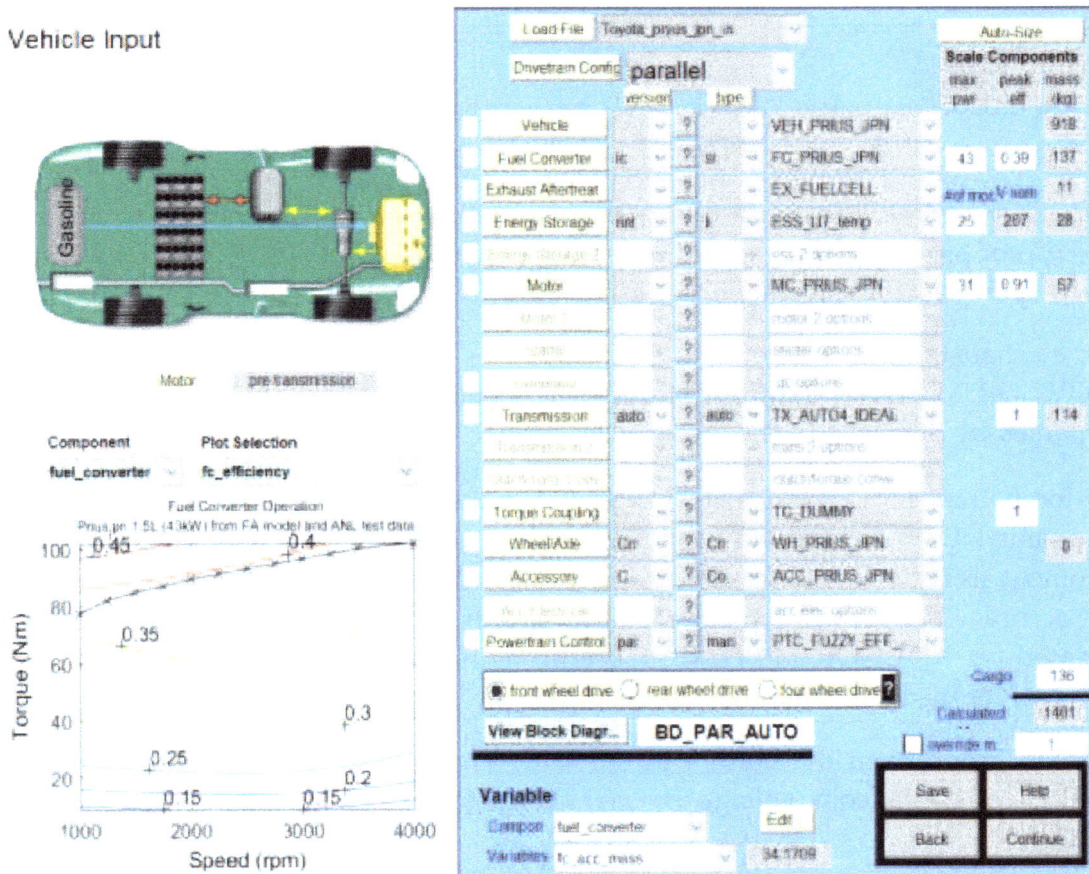

Figure 2. The setup of the Japanese Toyota Prius HEV' car in ADVISOR MATLAB platform under standard initial conditions (initial value of SOC of 70%).

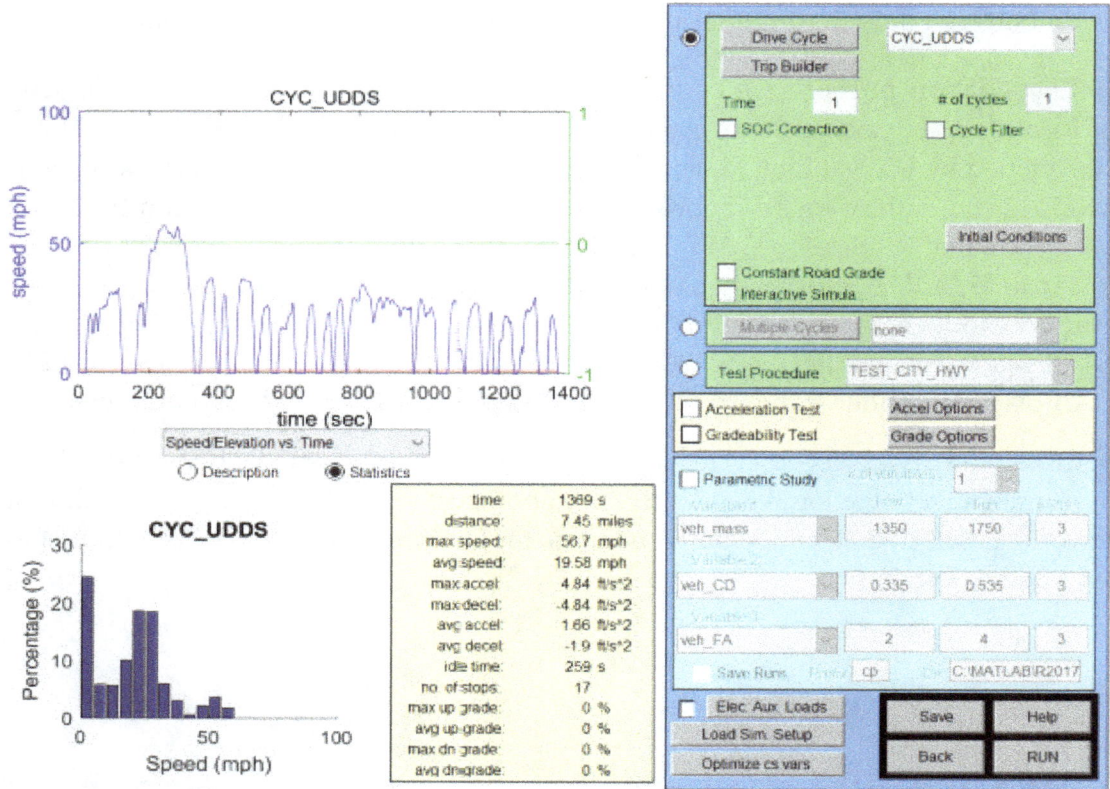

Figure 3. The UDDS cycle profile for Toyota Prius car speed test represented on the ADVISOR MATLAB platform.

Both, the driving UDDS cycle for car speed (*mph*) profile and its corresponding Li-Ion battery UDDS current profile are represented separately in the same ADVISOR MATLAB platform, as is shown in the first two corresponding top graphs from **Figure 4**.

Also, the exhaust gases emission and SOC curves, as a result to the same UDDS cycle test on the ADVISOR MATLAB platform, are shown in the last two bottom graphs of **Figure 4**, accompanied by details in the next section.

The Li-Ion battery EMC and the ADVISOR SOCs, for the same initial condition, SOC = 70% and the same UDDS cycle profile input current test, are implemented in a real-time MATLAB environment, as is shown on the same graph at the top of **Figure 5**. The SOC simulations reveal a great accuracy between Li-Ion battery EMC and its corresponding ADVISOR NREL model represented on ADVISOR MATLAB platform, as is shown in **Figure 4**, thus an expected confirmation of the EMC validation results.

The second graph from the bottom in the same **Figure 5** depicts the EMC battery output terminal DC instantaneous voltage $y = v(t)$, as a response to the UDDS cycle current input profile test.

The simulation results of Li-Ion battery EMC output voltage show a stabilization of the battery output voltage value at the end of UDDS cycle, i.e. after 1370 seconds.

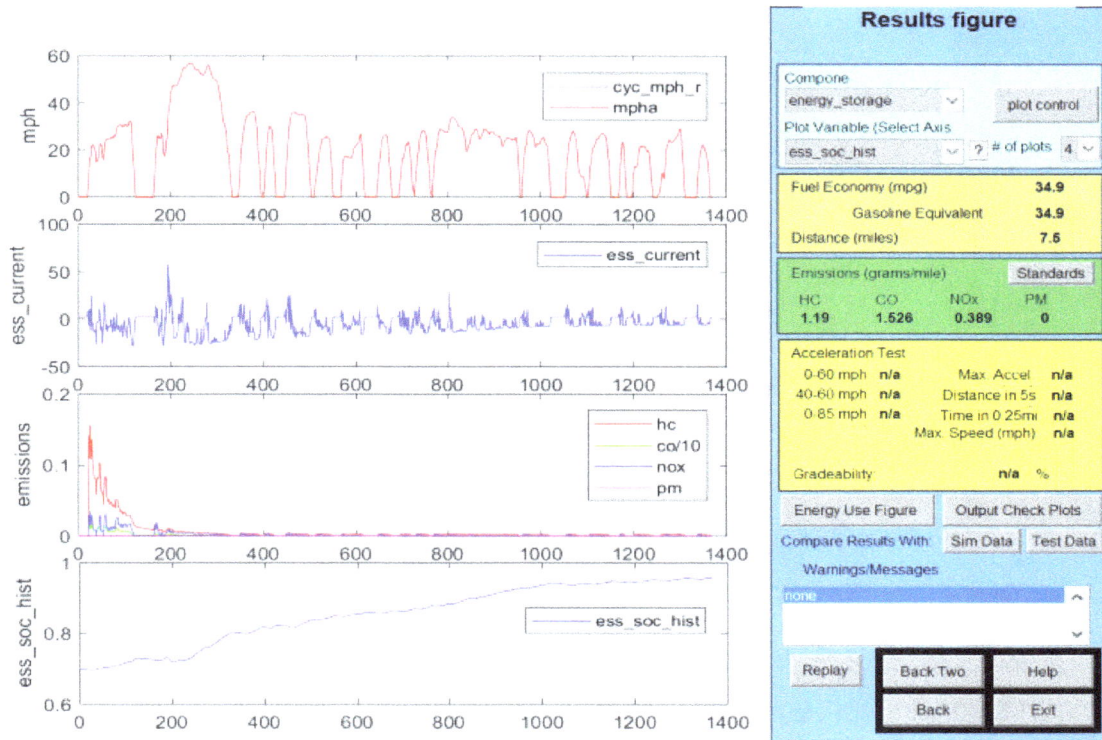

Figure 4. The corresponding car speed, current profile, exhaust gases emission, and SOC curves to UDDS cycle tests in ADVISOR MATLAB platform.

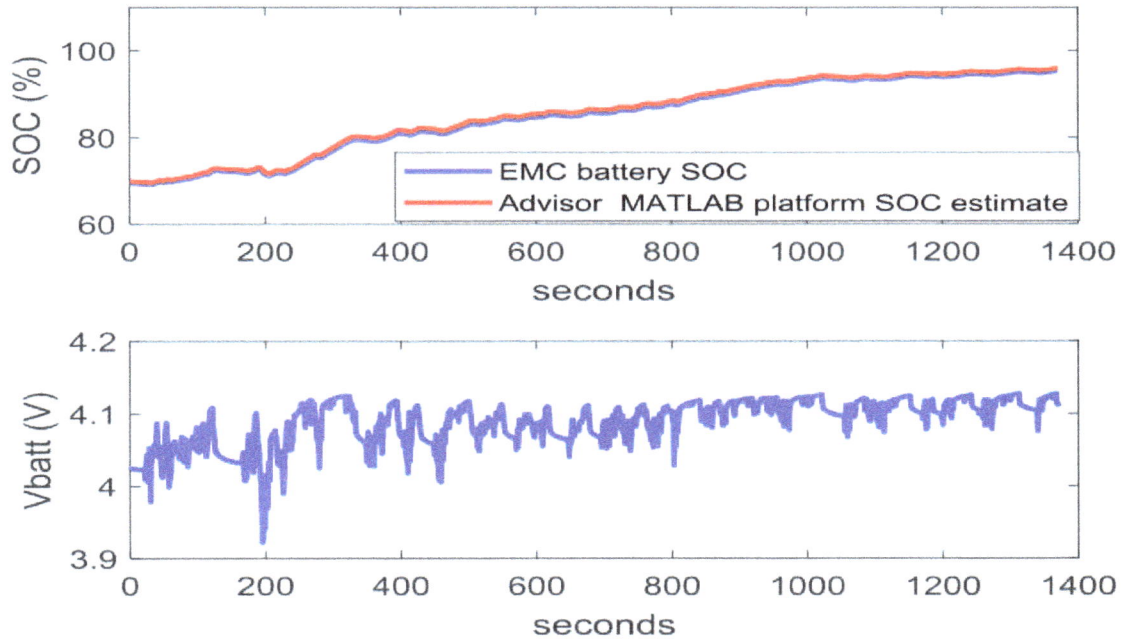

Figure 5. The SOC curves for Li-Ion battery EMC and NREL/6 Ah SAFT models in MATLAB R17a (the top graphs), and its corresponding EMC battery output terminal voltage, Vbatt (the second bottom graph).

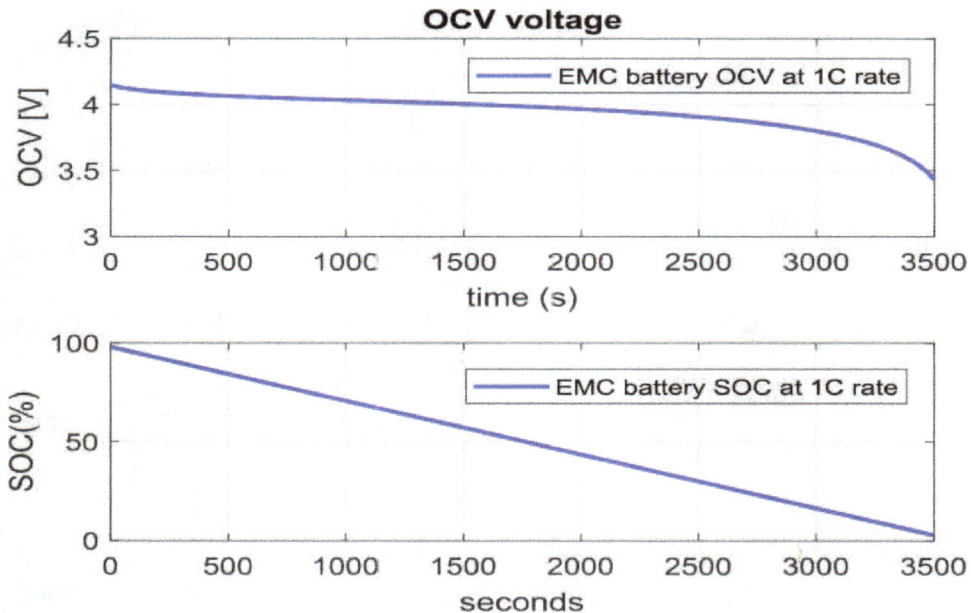

Figure 6. The Li-Ion battery EMC OCV curve during a complete discharging cycle at 1C-rate (top) and the corresponding battery SOC (bottom).

Likewise, the Li-Ion battery OCV for a discharging cycle at 1C-rate (i.e. 6 Ah × 1/h = 6A constant input discharging current) is shown in the top of **Figure 6**, and its corresponding SOC during the same discharging cycle is revealed on the bottom graph of the same **Figure 6** respectively.

3. Development and implementation in real-time of SOC Li-Ion battery estimators on MATLAB/SIMULINK platform

In this section we propose for Li-Ion battery EMC SOC estimation three nonlinear on-board real-time estimators integrated in BMS of HEV, based on Kalman Filter (KF) technique, specifically a nonlinear Gaussian Unscented Kalman Filter (UKF), a non-Gaussian nonlinear Particle Filter (PF), and a nonlinear observer estimator (NOE). The simulations results and a comprehensive performance analysis for each proposed SOC estimator are presented in the following subsections of this section.

3.1. Unscented Kalman filter real-time estimator design and robustness analysis

The main aim of this subsection is to build a nonlinear UKF SOC estimator, following the same design procedure described rigorously in [5]. We are motivated by some preliminary results obtained in our research, as you can see in [6, 7]. Technically, UKF estimator is based on the principle that one set of discrete sampled points parameterizes easily the mean and the covariance of a Gaussian random variable, as is stated in [5]. Moreover, the nonlinear estimator UKF yields an equivalent performance compared to a linear extended Kalman filter (EKF), well documented in [3, 4, 8], excluding the linearization steps required by EKF. In addition, the results of UKF real-time implementation for the majority of similar applications are encouraging, and it seems that the anticipated performance of this approach is slightly superior compared to EKF [3, 4, 8]. Furthermore, the nonlinear UKF SOC estimator can be extended to the applications where the distributions of the process and measurements noises are not Gaussian [19]. Concluding, the implementation simplicity and a great estimation accuracy of the proposed UKF SOC estimator recommend it as the most suitable estimator to be used in almost all similar applications. Explicitly, the nonlinear UKF estimator is an algorithm of *"predictor–corrector"* type applied to a nonlinear discrete - time systems, such as those described in [3–7]:

$$x(k+1) = f(x(k), u(k), k) + w(k) \tag{13}$$

$$y(k) = g(x(k), u(k), k) + v(k) \tag{14}$$

where $f(.)$, $g(.)$ are two nonlinear functions of system states and inputs; $w(k)$, $v(k)$ are zero-mean, uncorrelated process, and measurement Gaussian noise respectively. Since the noise injected in the state and output equations are randomly, also the system state vector and the output become random variables, having the mean and covariance matrices as a statistics. The nonlinear UKF SOC estimator has a *"predictor-corrector"* structure that propagates the mean and the covariance matrix of a Gaussian distribution for the random state variable $x(k)$ in a recursively way, in both prediction and correction phases, as is stated in [3–7]. The propagation of these first two moments is performed by using an unscented transformation (UT) to calculate the statistics of any random variable that undergoes a nonlinear transformation [5–7]. As is stated in the original and fundamental work [5], by means of this UT transformation *"a set of points, the so-called sigma points, are chosen such that their sample mean, and sample covariance matrix are $\overline{x}$ and P_{xx}, respectively"*. Also, *"the nonlinear function $g(.)$ is applied to each sigma points generating a "cloud" of transformed points with the mean $\overline{y}$ and the covariance matrix P_{yy}"*. Furthermore, since statistical convergence is not an issue, only a very small number of sample points is enough to capture high order information about the random state variable distribution. The different selection strategies of the sigma points and the UKF algorithm steps in a *"predictor-corrector"* structure, as well as its tuning parameters are well documented in the literature, and for details we recommend the fundamental work [5]. Since we follow the same design procedure steps for building a nonlinear UKF SOC estimator as in [5–7], we are focused only to the implementation aspects.

The simulation results of the real-time implementation of proposed UKF SOC nonlinear estimator in MATLAB R17a environment are shown in **Figures 7** and **8**. In **Figure 7** are presented the simulation results for EMC SOC true value versus EMC SOC-UKF and ADVISOR MATLAB platform estimated values.

Also, at the bottom of **Figure 7** is shown the EMC battery output terminal true value DC voltage versus EMC-UKF battery output terminal estimated DC voltage. The simulation results reveal an accurate SOC estimation values, and also a very good robustness of UKF estimator to the changes in initial SOC value (guess value, SOCinit = 20%). In **Figure 8** are shown the Li-Ion EMC polarization voltages versus Li-Ion EMC -UKF estimates, and in **Figure 9** is represented the robustness of EMC-UKF SOC nonlinear estimator to a gradual

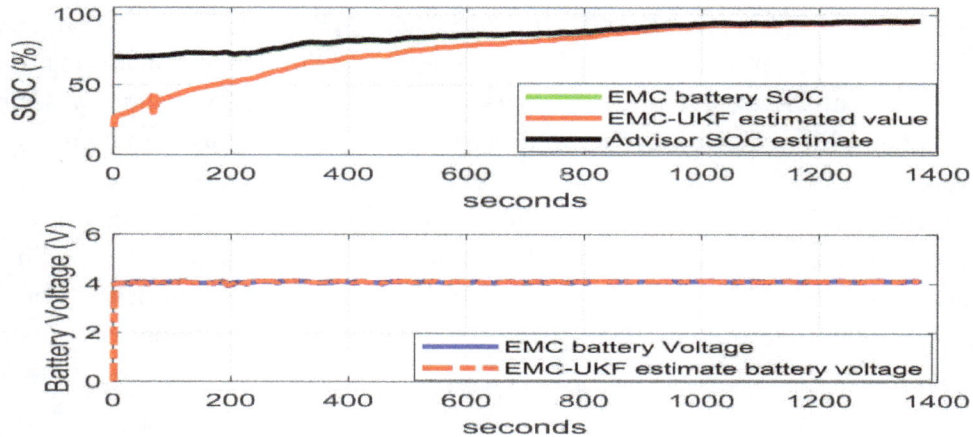

Figure 7. Li-Ion EMC SOC and output terminal battery DC voltage versus Li-Ion EMC-UKF and ADVISOR estimated values for an UDDS cycle current profile test.

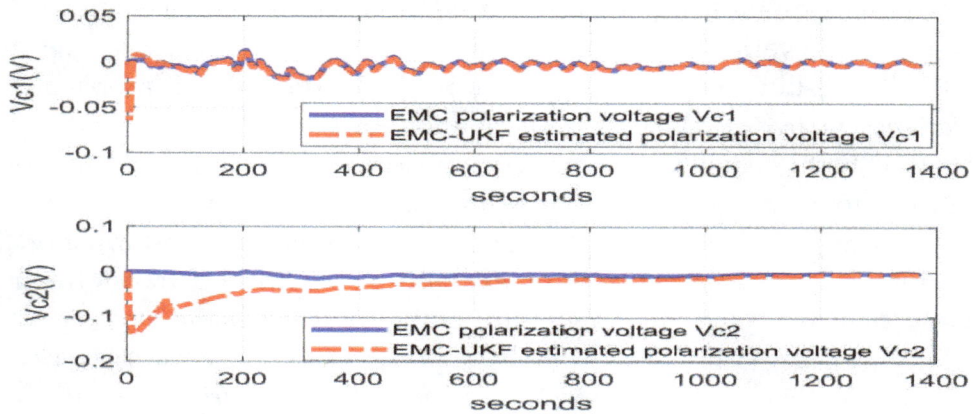

Figure 8. Li-Ion EMC polarization DC voltages versus Li-Ion EMC-UKF estimated values for an UDDS cycle current profile test.

increase in the internal resistance by 1.5 until 2 times of its initial value. The simulation results reveal a significant decrease in Li-Ion EMC UKF SOC estimator performance to an increase in internal resistance, but still remains convergent to EMC measurements after a long transient.

3.2. Particle filter real-time estimator design and robustness analysis

In this subsection we propose a real-time PF SOC nonlinear estimator with a similar" *prediction-corrector*" structure found to the nonlinear UKF SOC estimator described in the previous subsection 3.1. Consequently, is expected that the proposed nonlinear PF SOC estimator to update recursively an estimate of the state and to find the innovations driving a stochastic process given a sequence of observations, as is shown in detail in the original work [19]. In [19] is stated that the PF SOC estimator accomplishes this objective by a sequential Monte Carlo method (bootstrap filtering), a technique for implementing a recursive Bayesian filter by Monte Carlo simulations.

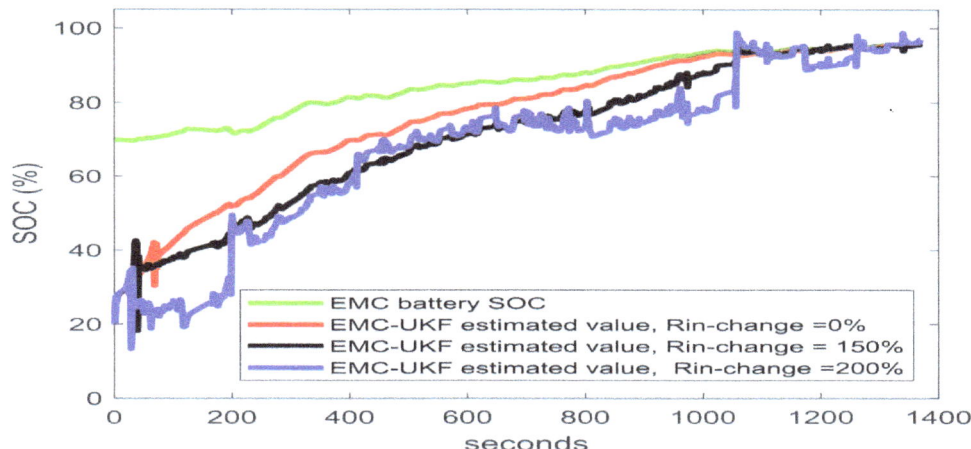

Figure 9. The robustness of Li-Ion EMC-UKF estimator to the changes in internal battery resistance for an UDDS cycle current profile test.

The process state estimates are used to predict and smooth the stochastic process, and with the innovations can be estimated the parameters of the linear or nonlinear dynamic model [19]. The basic idea of PF SOC estimator is that any probability distribution function (*pdf*) of a random variable can be represented as a set of samples (particles) as is described in [19], similar thru sigma points UKF SOC estimator technique developed in subsection 3.1 [5]. Each particle has one set of values for each process state variable. The novelty of this method is its ability to represent any arbitrary distribution, even if for non-Gaussian or multi-modal pdfs [5].

Compared to nonlinear UKF SOC estimator design, the nonlinear PF SOC estimator has almost a similar approach that does not require any local linearization technique, i.e. Jacobean matrices, or any rough functional approximation. Also, the PF can adjust the number of particles to match available computational resources, so a tradeoff between accuracy of estimate and required computation. Furthermore, it is computationally compliant even with complex, non-linear, non-Gaussian models, as a tradeoff between approximate solutions to complex nonlinear dynamic model versus exact solution to approximate dynamic model [6, 19]. In the Bayesian approach to dynamic state estimation the PF estimator attempts to construct the posterior probability function (pdf) of a random state variable based on available information, including the set of received measurements. Since the pdf represents all available statistical information, it can be considered as the complete solution to the optimal estimation problem. More information useful for design and implementation in real-time of a nonlinear PF estimator can be found in the fundamental work [19]. Since we follow the same design procedure steps to build and implement in real-time a nonlinear PF estimator, as is developed in [19], we are focused only on the implementation aspects. The simulation results in real-time MATLAB R17a environment are shown in **Figures 10** and **11**.

In **Figure 10** are shown the simulation results for EMC SOC true value versus EMC SOC-PF and ADVISOR MATLAB platform estimated values. The number of filter particles is set to 1000, a very influent tuning parameter for an accurate SOC estimation performance. At the bottom of **Figure 10** are shown the Li-Ion EMC battery output terminal true values of DC instantaneous voltage versus Li-Ion EMC battery output terminal DC voltage estimated by the

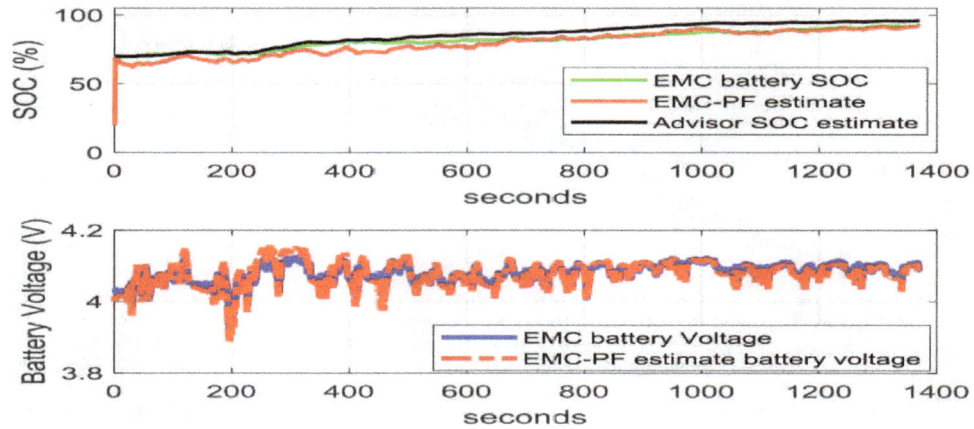

Figure 10. EMC SOC and output terminal voltage versus EMC-PF estimated values during UDDS cycle current profile test.

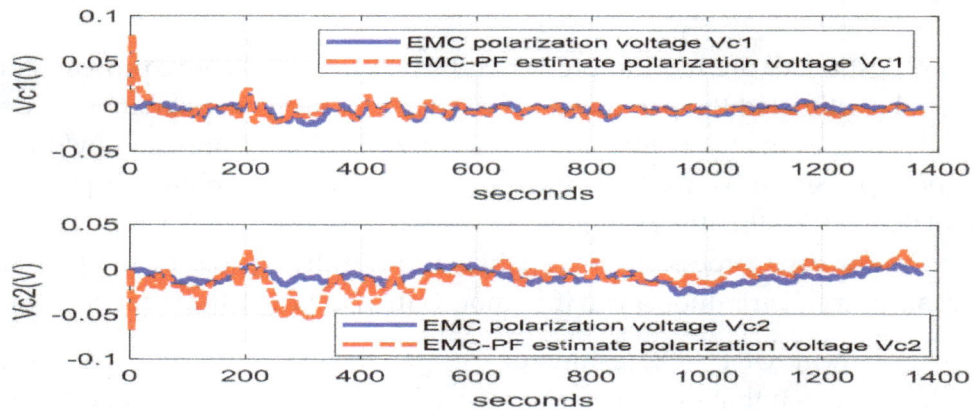

Figure 11. Li-Ion EMC polarization voltages versus Li-Ion EMC-PF estimated values during UDDS cycle current profile test.

nonlinear PF estimator. In **Figure 11** are shown the Li-Ion EMC polarization DC voltages versus EMC PF estimates. In **Figure 12** is shown the robustness test of Li-Ion EMC-PF SOC nonlinear estimator to a gradual increase in the internal battery resistance by same values considered for Li-Ion EMC-UKF SOC estimator.

The simulation results reveal a good robustness and convergence of Li-Ion EMC-PF estimator, but with a lot variance in the estimated values. Overall, the simulation results reveal a fast PF estimator convergence, a good SOC filtering, an accurate SOC estimation value, and also a very good robustness of PF estimator to big changes in the initial SOC value (guess value, SOCinit = 20%), and slightly slow behavior to an increase in internal resistance of the Li-Ion battery.

3.3. Nonlinear observer real-time estimator

In this subsection, a nonlinear observer SOC estimator (NOE) is under consideration. It is proposed to have more flexibility for a suitable choice of the best Li-Ion battery SOC estimator

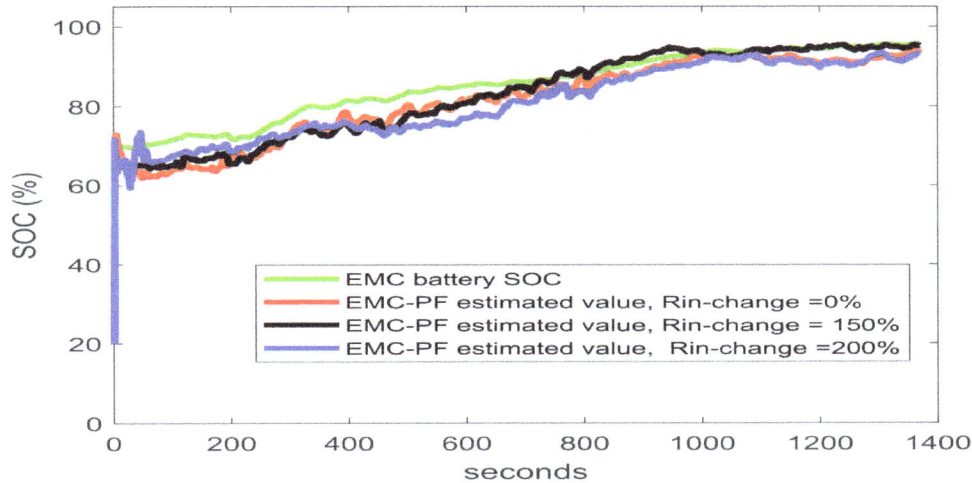

Figure 12. The robustness of EMC-PF estimator to the changes in internal battery resistance for an UDDS cycle current profile test.

amongst UKF, PF, and NOE SOC estimators. We follow the same design procedure steps for its design and implementation in a real-time MATLAB R2017a simulation environment as in [1]. The estimator design is based on an important information provided by the linear structure of the matrix Eq. (9).

According to this structure all three state variables $x_1(k), x_2(k), x_3(k) = SOC(k)$ change independently. Precisely, the nonlinear, linear and sliding mode observers are most applied in state estimation problems to eliminate state estimation error using deviation feedback, as is mentioned in [1]. Furthermore, for Li-Ion battery SOC estimation, most of existing observers are model based using for structure design the difference between the estimated value of battery output terminal DC instantaneous voltage and its corresponding measured DC voltage value multiplied by the observer gains to correct the dynamics of all estimated states, as follows:

$$
\begin{aligned}
&\widehat{x}(k+1) = A\,\widehat{x}\,(k) + Bu(k) + L_k(y(k) - \widehat{y}(k)), \\
&L_k = [l_{1k}\ l_{2k}\ l_{3k}]^T, \widehat{x}(k) = [\widehat{x}_1(k)\ \widehat{x}_2(k)\ \widehat{x}_3(k)]^T = \left[\widehat{V}_{C_1}\ \widehat{V}_{C_2}\ \widehat{SOC}\right]^T, \\
&\widehat{y}(k) = \widehat{V}_{batt} = VOC\left(\widehat{SOC}\right) - \widehat{x}_1(k) - \widehat{x}_2(k) - Ru(k) \\
&e_{x_1}(k) = e_{V_{C_1}}(k) = x_1(k) - \widehat{x}_1(k), e_{x_2}(k) = e_{V_{C_2}}(k) = x_2(k) - \widehat{x}_2(k), e_y = y(k) - \widehat{y}(k)
\end{aligned}
\tag{15}
$$

where the matrices A and B are the same as in Eq. (9), $\widehat{x}(k)$ is the estimate of the EMC states vector, and L_k is the observer gains vector (similar to the extended Luenberger observer for nonlinear systems). The particular structure of Li-Ion battery EMC reveals that the output estimation error e_y is mainly caused by an inaccurate SOC estimated value, as is stated also in [1]. Consequently, only the SOC state estimate from third discrete state Eq. (15) will be affected, i.e. the observer gains vector becomes:

$$l_{1k} = 0, l_{2k} = 0, l_{3k} \neq 0 \tag{16}$$

This outcome improves significant the NOE SOC estimation accuracy and simplifies the structural complexity of the proposed nonlinear observer estimator. The dynamics of the nonlinear observer estimation errors can be described by the following differential equations [1]:

$$
\begin{aligned}
e_{x_1}(k+1) &= \left(1 - \frac{T_s}{T_1}\right) e_{x_1}(k) \\
e_{x_2}(k+1) &= \left(1 - \frac{T_s}{T_2}\right) e_{x_2}(k) \\
e_{SOC}(k+1) &= e_{x_3}(k) = l_{3k} e_y(k)
\end{aligned}
\tag{17}
$$

In [1] is proved that all three states estimation errors described by the system of Eq. (17) converge asymptotically to zero in steady-state, and the observer gain for the new simplified structure is approximated by an adaptive law:

$$l_{3k} = l_{30} + \alpha e^{\beta(|e_y|)}, \ \ l_{30} > 0, \alpha < 0, \beta < 0 \tag{18}$$

that allows the value of l_{3k} to change dynamically according to the deviation between the measured battery output DC voltage and battery EMC output DC voltage. In Eq. (18), l_{30}, α and β are tuning parameters designed to adjust the adaptive property of l_{3k}. Amongst them, l_{30} determines the convergence rate of the proposed NOE at first "inaccurate" stage, the coefficients α and β are used to adjust observer gain l_{3k} when the SOC state estimation also reaches "accurate" stage, as is stated in [1].

Furthermore, three main assumptions are formulated in [1] to tune the values of EMC-NOE parameters l_{30}, α and β: (a) $l_{3k} \geq 0$ to ensure the stability of the proposed NOE; (b) if SOC state estimation error is large, the value of l_{3k} should be big enough to ensure a fast convergence rate; (c) if the voltage estimation error is small, the value of l_{3k} should be small enough to avoid SOC estimation "jitter" effect. By extensive simulations performed in a real-time MATLAB R2017a simulation environment the requirements (a), (b), (c) are met if the NOE parameters l_{30}, α and β are tuned for the following values: $l_{30} = 0.3$, $\alpha = -0.01$, and $\beta = -1$. The simulation results on the estimation performance of Li-Ion EMC-NOE are shown in **Figures 13** and **14**.

In **Figure 13** are shown the simulation results for Li-Ion EMC SOC true value versus Li-Ion EMC SOC-NOE and ADVISOR MATLAB platform estimated values. At the bottom of **Figure 13** are presented the Li-Ion EMC battery output terminal true values DC voltage versus Li-Ion EMC battery output terminal DC voltage estimated by the proposed Li-Ion EMC NOE SOC estimator. In **Figure 14** are displayed the Li-Ion EMC polarization DC voltages versus Li-Ion battery EMC-NOE estimates.

In **Figure 15** is depicted the robustness of Li-Ion EMC-NOE SOC estimator to an increase in internal battery Li-Ion resistance with the same values as used for the previous nonlinear estimators, EMC UKF and EMC PF respectively.

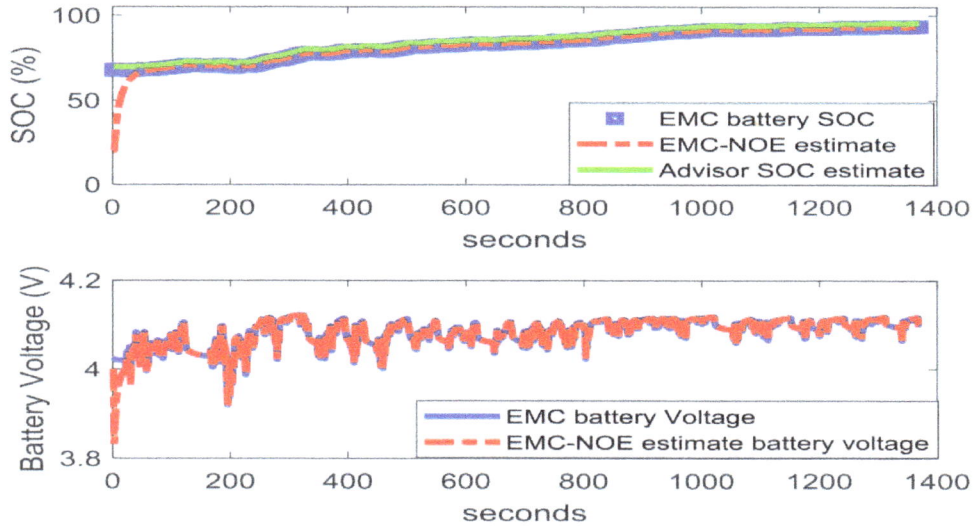

Figure 13. Li-Ion EMC SOC and output terminal DC voltage versus Li-Ion EMC-NOE estimated values during UDDS cycle current profile test.

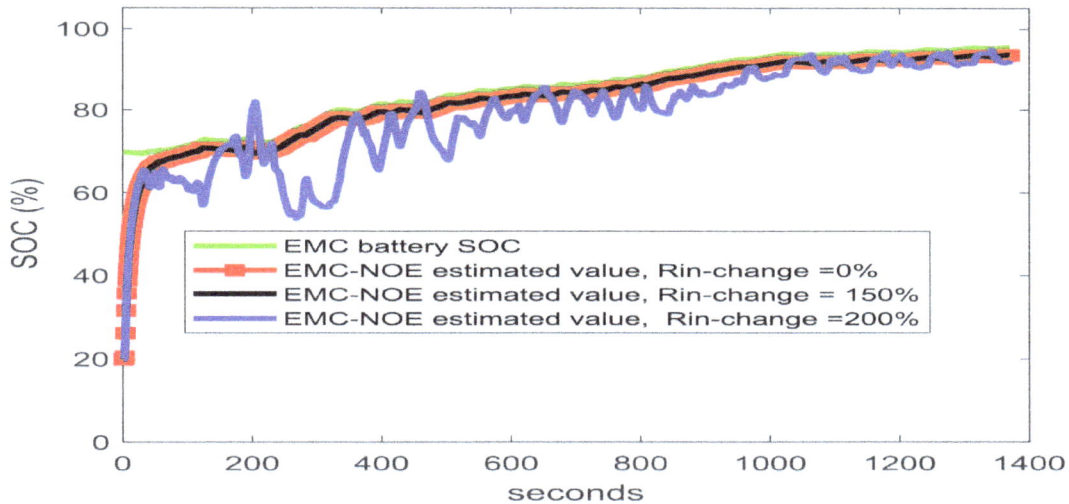

Figure 14. Li-Ion EMC-NOE polarization DC voltages versus Li-Ion EMC-NOE estimated values during UDDS cycle current profile test.

The simulation results in **Figure 15** reveal slightly slow robustness to an increase in internal resistance of Li-Ion battery compared to simulation results from **Figure 13**, but still the convergence is reached in a long transient with much variation in the SOC estimates. Overall, the simulation results from this section reveal a fast NOE estimator convergence, a good SOC filtering, an accurate SOC estimation value, and also a very good robustness of PF estimator to big changes in the initial SOC value (the same guess value as for UKF and PF, SOCinit = 20%), compared to a gradual degrade in SOC estimation performance to an increase in internal battery resistance.

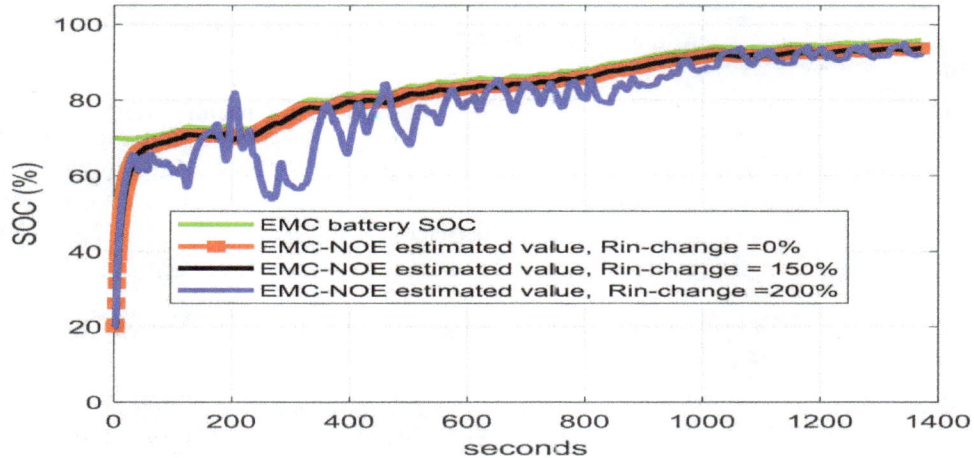

Figure 15. The robustness of Li-Ion EMC-NOE estimator to the changes in internal battery resistance for an UDDS cycle current profile test.

4. Real-time implementation of the Li-Ion battery SOC estimators on MATLAB/SIMULINK platform: Results comparison

For comparison purpose, we represent on the same graph the SOC estimates of all three real-time nonlinear estimators (UKF, PF, and NOE) versus EMC SOC true value and ADVISOR MATLAB SOC estimate, as are revealed in **Figures 7, 10** and **13**. The simulation results in **Figure 16** disclose a precious information about the convergence of all three proposed estimators to Li-Ion EMC SOC true values for the proposed Li-Ion battery EMC. Furthermore, a useful benchmark is built in terms of the statistics errors between the states estimates and the corresponding model states values, such as root mean square error (RMSE), mean square error (MSE) and mean absolute error (MAE), defined in [6]. They are very easy to be computed in MATLAB R2017a, and the results are shown in **Table 1**. The MSE is a measure of how close the estimates values fit to model "true" values. The squaring is done so negative values do not cancel positive values. The smaller the MSE, the closer the fit of the estimates values is to the model "true" values. The RMSE is just the square root of the MSE [6].

$RMSE = \sqrt{\dfrac{\sum_{i=1}^{i=N} (x_{est}(i) - x_{model}(i))^2}{N}} = \sqrt{MSE}$ The RMSE is probably the most easily interpreted statistic, since it has the same units as the model states. Similar as MSE, lower RMSE the state estimates fit better the model states, i.e. battery SOC and the battery polarization voltages. The MAE statistic is helpful to determine the accuracy of the Li-Ion battery EMC-UKF, EMC-PF and EMC-NOE SOC estimates with respect to the "true" model values. It is usually similar in magnitude to RMSE, but slightly smaller, and has the same units as the model states data set.

$MAE = \dfrac{\sum_{i=1}^{i=N} (|x_{est}(i) - x_{model}(i)|)}{N}$ The statistics of errors from benchmark reveal that EMC nonlinear observer estimator (EMC-NOE) outperforms in terms of all statistics (RMSE, MSE, MAE) the

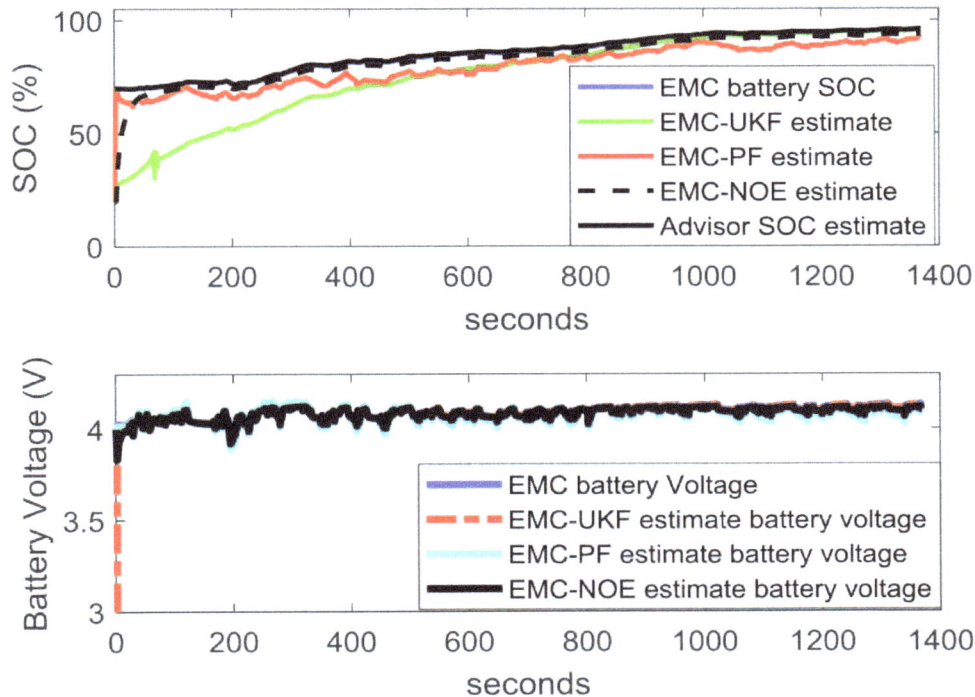

Figure 16. EMC SOCs and output terminal voltages versus EMC, EMC-UKF, EMC-PF, and EMC-NOE estimated values during UDDS cycle current profile test.

UKF estimator			PF estimator			NOE		
MAE	MSE	RMSE	MAE	MSE	RMSE	MAE	MSE	RMSE
0.0915	26.1030	0.1380	0.0572	4.9724	0.0602	0.0247	2.4584	0.0424

Table 1. Statistics on SOC estimation performance of the proposed nonlinear estimators UKF, PF and NOE - Benchmark.

UKF and PF estimators. Consequently, for this kind of applications the EMC-NOE SOC estimator is the most suitable estimator compare to UKF and PF estimators.

5. Conclusions

The main contribution of this research paper is the design and implementation in real time of a three robust nonlinear estimators, namely UKF, PF and NOE, capable to estimate with high accuracy and robustness the Li-Ion battery SOC based on a simple battery 2R EMC without disturbance uncertainties. The simulation results obtained in a real-time MATLAB simulation environment reveal that amongst all three proposed nonlinear SOC estimators the EMC NOE is a most suitable alternative to SOC UKF and PF estimators for this kind of application. The number of tuning parameters for SOC EMC NOE is much smaller than for UKF and PF estimators. The proposed SOC EMC NOE proves its effectiveness in terms of implementation simplicity, Li-Ion battery SOC estimation accuracy and robustness. Therefore, it can be considered as one of the most suitable nonlinear estimator, and a feasible alternative to UKF and PF estimators.

Conflict of interest

The authors declare that there is no conflict of interests regarding the publication of this paper.

Author details

Roxana-Elena Tudoroiu[1], Mohammed Zaheeruddin[2], Sorin-Mihai Radu[1] and Nicolae Tudoroiu[3]*

*Address all correspondence to: ntudoroiu@gmail.com

1 University of Petrosani, Petrosani, Romania

2 University Concordia from Montreal, Montreal, Canada

3 John Abbott College, Saint-Anne-de-Bellevue, Canada

References

[1] Xia B, Zheng W, Zhang R, Lao Z, Sun Z. Mint: A novel observer for Lithium-ion battery state of charge estimation in electric vehicles based on a second-order equivalent circuit model. Energies. 2017;**10**(8):1150. DOI: 10.3390/en10081150. Available from: http://www.mdpi.com/1996-1073/10/8/1150/htm. [Accessed: 2017-01-21]

[2] Young K, Wang C, Wang LY, Strunz K. Electric vehicle battery technologies–chapter 2. In: Garcia-Valle R, Lopes JAP, editors. Electric Vehicle Integration into Modern Power Networks. 1st, 9, and 325 ed. New-York, USA: Springer Link: Springer-Verlag; 2013. pp. 15-26. DOI: 10.1007/978-1-4614-0134-6.ch.2

[3] Plett GL. Extended Kalman filtering for battery management systems of LiPB-based HEV battery packs - part 1. In: Modeling and Identification, Power Sources. Vol. 134. Amsterdam, Netherlands: Elsevier B.V.; 2004. pp. 252-261. DOI: 10.1016/j.jpowsour.2004.02.031

[4] Plett GL. Extended Kalman filtering for battery management systems of LiPB-based HEV battery packs - part 2. In: Modeling and Identification, Power Sources. Vol. 134. Amsterdam, Netherlands: Elsevier B.V.; 2004. pp. 262-276. DOI: 10.1016/j.jpowsour.2004.02.032

[5] Simon JJ, Uhlmann JK. A new extension of the Kalman filter to nonlinear systems. [Internet]. In: Process of AeroSense, 11th International Symposium on Aerospace/Defense Sensing, Simulation and Controls; 1997. Available from: https://people.eecs.berkeley.edu/~pabbeel/cs287-fa09/readings/JulierUhlmann-UKF.pdf. [Accessed: 2017-01-21].

[6] Tudoroiu N, Radu SM, Tudoroiu E-R. Improving Nonlinear State Estimation Techniques by Hybrid Structures. 1st ed. Saarbrucken, Germany: LAMBERT Academic Publishing; 2017. p. 56 ISBN: 978-3-330-04418-0

[7] Tudoroiu N, Zaheeruddin M, Cretu V, Tudoroiu E-R. IMM-UKF versus frequency analysis [past and present]. IEEE Industrial Electronics Magazine. 2010;4(3):7-18. DOI: 10.1109/MIE.2010.937937

[8] Farag M. Lithium-ion batteries. In: Modeling and state of charge estimation (Thesis). Ontario, Canada: McMaster University of Hamilton; 2013. p. 169

[9] Johnson VH. Battery performance models in ADVISOR. Journal of Power Sources, Elsevier Science B.V. Publishing. 2001;110:321-329

[10] Tremblay O, Dessaint L-A. A generic battery model for the dynamic simulation of hybrid electric vehicles. IEEE Xplore. 2007:284-289. DOI: 10.1109/VPPC.2007.4544139

[11] Xu J, Cao B. Battery management system for electric drive vehicles-modeling state estimation and balancing - chapter 4. In: Chomat M, editor. New Applications of Electric Drives. Croatia: Intech; 2015. pp. 87-113. DOI: 10.5772/61609

[12] Nordelof A, Messagie M, Tilmann A-M, Soderman ML, Mierlo JV. Environmental impacts of hybrid, plug-in hybrid, and battery electric vehicles. Journal of Life Cycle assessment. 2014;19(11):1866-1890. DOI: 10.1007/s11367-014-0788-0

[13] Xing Y, Ma EWM, Tsui KL, Pecht M. Battery management systems in electric and hybrid vehicles. Energies. 2011;4:1840-1857. DOI: 10.3390/en4111840

[14] Battery Management System. [Internet]. Available from the web site: https://www.nasa.gov/centers/johnson/techtransfer/technology/MSC-24466-1_Batt-Mgmt-Sys.html. [Accessed: 2018-02-10]

[15] Prophet G. 12-Cell Li-Ion Battery Monitor for EV/HEV Protection. 2016. [Internet]. Available from the web site: http://www.eenewspower.com/news/12-cell-li-ion-battery-monitor-evhev-protection. [Accessed: 2018-02-10]

[16] Jiang J, Zhang C. Fundamentals and Applications of Lithium-Ion Batteries in Electric Drive Vehicles. 1st ed. Singapore: John Wiley & Sons Singapore Pvt Ltd. p. 300; ISBN: 978-1-118-41478-1

[17] Dey S, Mohon S, Pisu P, Ayalew B. Sensor detection, isolation, and estimation in Lithium-ion batteries. IEEE Transactions on Control Systems Technology. 2016;24(6):2141-2149. DOI: 10.1109/TCST.2016.2538200

[18] Singh A, Izadian A, Anwar S. Nonlinear model based fault detection of Lithium ion battery using multiple model adaptive estimation. Proceedings of IFAC, Cape Town. 2014;47(3):8546-8551. DOI: https://doi.org/10.3182/20140824-6-ZA-1003.00711

[19] Arulampalam MS, Maskell S, Gordon N, Clapp T. A tutorial on particle filters for online nonlinear/non-Gaussian Bayesian tracking. IEEE Transactions on Signal Processing. 2002;50(2):174-188. DOI: 10.1109/78.978374

2

Outer Rotor SRM Design for Electric Vehicle without Reducer via Speed-Up Evolutionary Algorithm

Zeki Omaç, Mehmet Polat, Mustafa Kaya,
Eyyüp Öksüztepe, Haluk Eren, Merve Yıldırım and
Hasan Kürüm

Abstract

Reducers utilized in automotive industry provide motor to run in most effective region and transmission output torque to increase. However, they cause mass and cost to increase and also efficiency to decrease due to mechanical losses. The aim of this study is to design a direct drive motor (outer rotor switched reluctance motor (OR-SRM)) without reducer resulting in enhanced efficiency for electric vehicle (EV). To estimate dimension and electrical parameters of OR-SRM, mathematical equations are originally derived from its geometry. Considering the constraints of package size and outer diameter, all the dimension parameters of the motor are optimized via multi-objective genetic algorithm (MOGA) to get the desired efficiency and torque. In order to validate the results in the proposed approach, OR-SRM is modeled by Maxwell 3D using optimized dimension parameters. In-wheel OR-SRM with 18/12 poles (30 kW) is manufactured to employ it in an EV. Theoretical results are compared to experimental results. It can be concluded that the results are satisfactory.

Keywords: outer rotor SRM design, nonreducer, electric vehicle, multi-objective genetic algorithm optimization

1. Introduction

Nonreducer structure of the direct drive systems provides avoiding extra mass and cost. They also enable to enhance energy efficiency due to decreasing mechanical losses. It is still a matter of debate how much benefit we can get with in-wheel motors. From a research perspective, authorities are very optimistic, but, economically, it does not make sense for commercial vehicles right now. Coordinating the in-wheel motors requires complex control systems, and it is not so hard to foresee that in-wheel technology takes not too long to be competitive.

Developers deserve to design motors with simple structure, high power density, and low cost. Permanent magnet synchronous motors (PMSMs) match aforementioned requirements, and they have been commonly employed in commercial EVs. PMSMs are widely employed in EVs due to their high power density and efficiency. Another motor type used in EVs is PM-assisted SynRM motors. They have important advantages such as low rotor losses, high saliency ratio (Lq/Ld), high torque, and high power factor. Furthermore, brushless DC motors are used in EVs due to having high torque at low speed and a proper torque-speed characteristic [1]. However, there are some drawbacks for these three motor types. The centrifugal force in the motor causes the stress in the rotor. This stress should be calculated and minimized by design optimizations [2, 3]. Besides, the magnets may be demagnetized due to reverse magnetic field and high temperature. They also cause uncontrollable regenerative operation owing to constant magnet flux existence. And also, magnet cost is considerable. SRMs are employed by various application areas such as aircrafts, EVs, marine propulsion systems, linear drives, mining drives, handheld tools, home utilities, etc. because of its inherent modularity and fault tolerance [4]. The motor behavior does not depend on the temperature due to not having the magnets [5]. There are many advantages of SRMs for EV such as not having brush, collector, magnets, and rotor consisting of only silicone sheets. It results in low production cost [6]. There is less maintenance requirement than the other electric machines. Not having magnet provides the motor to be operated at high speed. Rotor copper losses do not exist due to none rotor winding. Therefore, rotor temperature is lower than that of other motor types, and motor can be easily cooled. Low inertia is significant in variable reference speed applications to get fast response for reference speed. The rotor of SRM has also lower inertia than the other motors, and the efficiency of SRM is over 95% [7]. Once one of the phases fails, the motor continues to operate due to not any connection between the phases [8]. That property makes it a very preferable choice for aircraft applications [9]. Conversely, SRM needs both a motor driver and a rotor position sensor [1, 10]. On the account of all features, it is seen that SRM is a proper motor type for EV applications and it may become promising for the near future.

In this study, we have investigated OR-SRM with outer rotor structure unlike conventional SRMs. In-wheel SRM is more preferable due to losses and weight of driveline rather non-in-wheel scheme. In spite of increasing the system complexity, coordinating each motor enables numerous advantages with regard to safety and efficiency. For example, the drive structure of the car could be flexibly reorganized in front, rear, or four-wheel considering the road conditions.

1.1. Problem statement

The essential forces influencing movement of a vehicle are in three folds, which are a force causing gravity as climbing up a ramp with specified slope, aerodynamic force varying with square of vehicle velocity, and required force of a vehicle having specified mass to accelerate. These are vehicle dynamics employed in this study. A motor for an EV should be designed both to directly (nonreducer) generate a torque satisfying aforementioned forces and to have a size fitting specified tire. A motor having adequately small size to fit in a tire needs high currents to keep generating a continuous high torque. High-efficient electric motor should be designed to increase the range of EV.

Electric motor losses including copper, friction, windage, and iron losses play an important role in energy efficiency. Energy efficiency is directly proportional to the range, which affects the range of EV. Predominantly, one of the losses is copper losses. To reduce these, battery voltage is increased and/or winding resistance is decreased. When battery voltage is increased for a motor having same power, the current reduces. Therefore, the copper losses will decrease. However, increasing battery voltage is restricted due to motor driver and personal safety. Winding resistance can be reduced as conductor cross section is enlarged. It may not be possible to use a very large conductor cross section residing in the motor both having a size to fit in a tire and generating required torque. Thus, a maximum conductor cross section for winding should be selected such that it can fit in splines of the motor.

The torque of SRM is directly proportional to both square of the motor current and derivative of the motor phase inductance with respect to rotor position. Variation of motor phase inductance depends on dimension parameters of the motor. Therefore, these parameters should be optimized to maximize torque and efficiency of the motor. The parameters of OR-SRM consist of stator-rotor pole arc angles and pole heights, inner and outer yoke lengths, and outer diameter of the stator. Optimization of parameters using conventional or linear methods takes long time, or it might be impossible, since solution space expands as the number of parameters to optimize increases. When the dimension parameters of the OR-SRM are optimized, magnetic circuit equations should be formulized to decide whether objective function is feasible for dimension parameters. As a motor is constructed with the mentioned design considerations, problem statement can be provided in brief as follows:

- How can a motor be designed such that its torque should directly satisfy the forces on a vehicle having a weight of nearly 2 tons and also it should have a dimension to fit the tire?

- While a motor having specified constraints generates required high torque, how the heat dissipation problem could be solved?

- How should an efficient motor be designed to increase range of the car?

- There are a lot of dimension parameters effecting efficiency and torque of the reluctance motor. Which parameters can be optimized?

- How the magnetic circuit equations associated with the dimension parameters of OR-SRM can be derived to employ them in the optimization problem?

1.2. Related work

Designing a high torque motor is deserved for direct traction EVs. Thus, diameter and package length of the machine should be configured such that designed motor can fit into each wheel. Relation between torque and dimensions including diameter and length of SRM was explained in detail [11–13]. In a research on designing SRM [14], the design and analysis with FEM of SRMs were studied, and the motor characteristics such as flux density, torque, current, and inductance values were investigated. Another paper presented a design, analytical verification, implementation, and testing of a 250 W, 200 rpm, 36 V, 15 Nm SR hub motor. Infolytica Magnet 6.24.1 software was used for verifying winding inductance and torque. According to the authors, the proposed motor with exterior motor structure could be easily incorporated within the wheels of a low-cost, low-weight EV [15]. In [16, 17], theoretical calculations of SRM and numerical analysis by 2D FEM were carried out. Also, dynamic behavior of a 12/8 SRM and the effect of motor characteristics were investigated. In [18], design and analysis of a SRM for next-generation hybrid vehicle without PM materials were realized. Maximum torque and efficiency of SRM were investigated for 6/4, 8/6, 12/8, and 18/12 SRMs as varying stator taper angles. Authors in [19] investigated a prototype machine of a SRM with respect to an IPMSM in a HEV, in terms of torque density, efficiency, and torque-speed range. Static and light load tests were realized for these machines. The authors in [20] presented a novel multi-objective optimization method based on a genetic fuzzy algorithm. Optimum design of SRM was obtained by this method in terms of two objective functions, which are high efficiency and low torque ripple. Optimum design approach for a two-phase SRM drive via genetic algorithm (GA) was proposed in [21]. Three GA loops work to optimize the lamination design and to meet the requirements for the target application while simultaneously fine-tuning the control parameters. A dynamic simulator based on an analytical expression of magnetizing curves and a geometric flux-tube-based nonlinear magnetic analysis was developed specifically for this class of motor. The experimental results of two-phase 8/6 prototype manufactured were validated with the optimized design. 8/6 SRM was designed by Taguchi optimization method for applications requiring fast actuation [22]. This was accomplished by two simultaneous optimizations of SRM which were optimization of motor torque and the torque per rotor inertia. Two orthogonal arrays (OA) were used for leading the design of experiments (DOE), and FEA was used for computing the performance of the motor designs generated by the Taguchi DOE. Authors claimed that this method provided a high electromagnetic torque to inertia ratio required for high rates of mechanical acceleration and robustness. Analysis and nonlinear optimization of four phases 8/14 SRM were accomplished by a combination of MATLAB and 2D FEM Software Ansoft Maxwell [23]. Here, torque ripple was minimized; all the factors including torque, torque quality, torque density, and losses were maximized by using GA. Reference [24] suggested an optimization method based on GA for an efficient design of a SRM. Some parameters such as stator and rotor pole arc, rotor diameter, and stack length were determined in this paper. Magnetic field analysis was done using FEA-based CAD package, and optimum results were obtained by GA. MOGA function including efficiency and torque ripple is presented to obtain optimum design parameters that are turn-on and turn-off angles and limited current in switching elements [25]. Improved magnetic equivalent circuit (IMEC) method was used to predict the performance of SRM and to generate the input–output data. Hence, the optimal fuzzy rules were determined. M.C. Costa and his colleagues realized optimization of 4/6 SRM to estimate the most significant parameters maximizing the magnetic torque of the motor and diffuse elements response surface [26]. The diffuse element method was also used to build a response surface representing an approximation of the real objective function. This study provided reduction of optimization parameters, reducing the computation time and increasing of the magnetic torque. In another paper, optimum design of pole shapes of SRM with three-phase 6/4 pole was accomplished by ordinary kriging model and GA to minimize torque ripple [27]. As the kriging model with the uniform sampling was used to construct a response surface, GA finds an optimal point of the approximate objective function. The proposed method was verified with high computation efficiency in terms of reducing the torque ripple. Cong Ma and his friends [28] presented multi-objective optimization of SRM based on a combination of DOE and particle swarm optimization. Maximizing torque per active mass and efficiency and minimizing torque ripple were provided by this method. Authors in [29] carried out the design and analysis of an in-wheel OR-SRM with 8/6 poles for electric bus applications. Derivation of the output power equation as a function of the motor dimensions and parameters was handled. The results of the developed machine obtained by FEA were compared with that of a conventional SRM. In this study, torque ripple decreased and motor efficiency considerably increased. Multi-objective optimization design of in-wheel SRMs in EV was realized in [30]. An optimization function was developed to maximize average torque, average torque per copper loss, and average torque per motor lamination volume. The stator and rotor pole arc angles were taken as the optimized variables. A proto-type of the optimally designed in-wheel SRM for EVs was designed. In [31], comparison of outer and inner rotor SRM was carried out. The efficiency and torque per ampere ratio of outer rotor machine were 92% and 12.8 Nm/A, respectively, and for inner rotor machine in all the cases, they were 90.7% and 12 Nm/A, respectively. The paper showed that the outer rotor machine was suitable for applications requiring high torque density and efficiency. A paper [32] presented the design of a SRM having torque, power, speed-range, and efficiency values competitive to those of the interior IPMSM employed in the 2009 Toyota Prius. The outer diameter and axial length were the same as those of the IPMSM. The simulation results showed that the shaft output power of the SRM enhanced to 1.6 times than the power of the IPMSM at high speed, while its current density and weight were increased by about 15–25%.

1.3. Contribution and proposed approach

We can find numerous studies in literature relevant to SRM design for electric car. However, in the course of designing direct drive motor for EVs, there is no study considering the following factors all together that are aforementioned vehicle dynamics, the constraints including outer diameter and package length of the motor, and five independent variables of motor dimension employed by evolution algorithm even if some of studies exist handling them individually. In this study for the subject EV, a direct drive OR-SRM mounted in wheel is designed and manufactured. Considering vehicle dynamics as designing SRM, optimum design parameters of OR-SRM are estimated by GA to raise efficiency and satisfy the deserved torque. Furthermore, OR-SRM magnetic circuit equations are derived by using seven base flux paths, which are utilized in calculation of motor efficiency and its torque. Nonlinear solution is carried out, involving B-H characteristics of silicon sheet in the equations. The multi-objective optimization problem is solved by accounting some of constraints such as package length, stator outer diameter fitting into the rim, and well-known SRM design considerations. In the present study, multiparameters including stator and rotor pole arc angles, their yoke lengths, and inner diameter of rotor are optimized using objective functions such that the highest efficiency for deserved torque is obtained.

1.4. Layout

Section 2 describes stages of an 18/12 OR-SRM design. Besides, determination of output torque, stator and rotor pole arcs, and calculation of other parameters are comprehensively explained in this section. In Section 3, the analysis of motor via Maxwell 3D software package is carried out. MATLAB analytical solution results are compared to Maxwell 3D results. Experimental results and discussions are provided, and comparison of Maxwell 3D and experimental results are performed in Section 4. Conclusions are given at the end.

2. Preliminaries of the system

In this study, a direct drive motor for EV has been designed and manufactured by using evolutionary algorithm. There are several factors reducing torque ripple of SRMs such as number of phases, pole numbers of stator and rotor, and trigger angle of phases. However, the increasing number of phases causes complexity of control and drive circuit system and also production cost to rise. Motor torque is estimated via specified factors affecting the vehicle performance that are acceleration, aerodynamic forces, and road slope. Specified rim size is selected such that it can fit the motor providing the required torque. And, the voltage and current values are selected considering nominal speed of the vehicle. Furthermore, OR-SRM nonlinear magnetic equations regarding the motor dimension are obtained. Considering the constraints of rim size and OR-SRM base speed, the motor design is performed by genetic algorithm. Eventually, the results of Maxwell 3D software are obtained with the objective of validation. The system diagram of the proposed approach is given in **Figure 1**, and its expansion is provided in the rest of the present section.

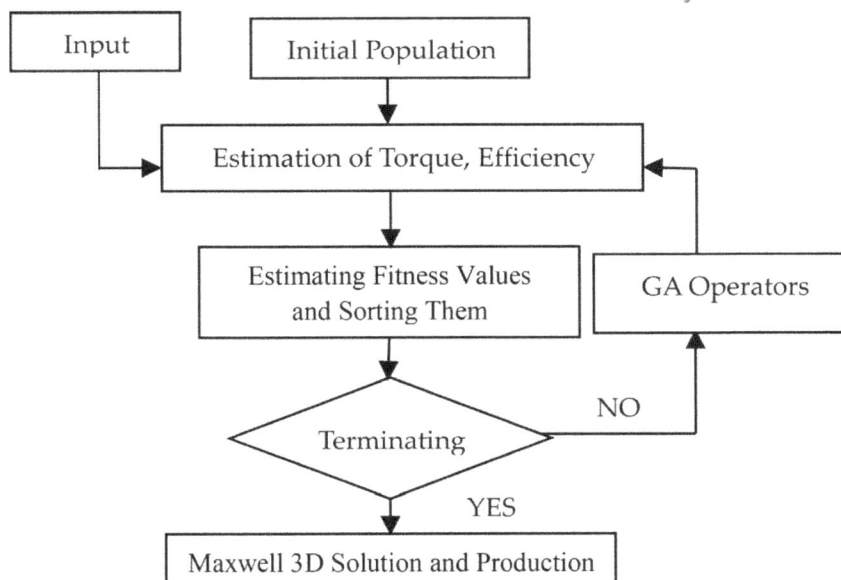

Figure 1. The system diagram of the proposed approach.

2.1. Estimation of torque generated by OR-SRM

The electric motors to be manufactured should drive the car to match the following conditions:

1. The EV speed should be up to 120 km/h (aerodynamic force).

2. It should be able to accelerate to 100 km/h in 10 s (aerodynamic and acceleration forces).

3. It should be able to climb 6% slope at 80 km/h constant speed (aerodynamic and slope forces).

When the torque of EV is calculated for three conditions given as above, the highest torque can be obtained in condition 2. If the motors satisfy condition 2, it means that they already satisfy the other conditions. Therefore, the torque estimation should be performed considering condition 2. In this situation, the torque equation can be estimated as

$$T_r = \left(m_v \cdot a_v + \frac{1}{2} \cdot c_w \cdot \delta_a \cdot s_f \cdot v^2 + c_r \cdot m_v \cdot g \right) \cdot r \tag{1}$$

where T_r represents the total torque to be generated by the motor, m_v refers to EV mass, a_v is the acceleration of EV, c_w is aerodynamic coefficient of the EV, δ_a is air density, s_f is perpendicular cross section of the EV, v is the speed of the EV, c_r is the rolling coefficient, g (9.81 m/s^2) is acceleration of gravity, and r is the radius of the EV wheel. These specified EV parameters and estimated torque value for each motor are shown in **Table 1**. Also, the range of EV on a not sloping road which has 90 km/h speed is approximately calculated as 230 km.

2.2. Deriving equations of OR-SRM electric and dimension parameters

Considering the aforementioned number of poles, outer diameter, package size, and air gap, electrical and dimensional parameters are estimated. Dimension parameters of OR-SRM in the equations are shown in **Figure 2**.

B-H characteristic of magnetic sheet plays a crucial role. Therefore, B-H characteristic is considered by catalog data, as seen in **Figure 3**. To put into practice the design of reluctance motor, magnetic induction density is desired not to excess knee point of B-H curve under nominal conditions. Therefore, the following situations are considered by:

1. If pole angle of rotor is greater than that of stator, knee point is taken as maximum rate of magnetic induction density.

2. If pole angle of stator is greater than that of rotor, maximum value of magnetic induction density is specified under knee point, such that avoiding excessive saturation and this value can be obtained considering proportion surface fields of stator and rotor pole.

m_v (kg)	S_f (m^2)	δ_a (m^3/s)	c_w	c_w	r (m)	Estimated torque for each motor (Nm)
1500	1.5	1.255	0.3	0.03	0.37	451.675

Table 1. Specified EV parameters and estimated torque value for each motor.

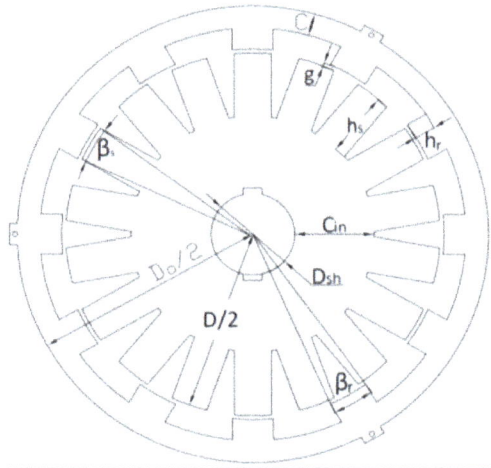

Figure 2. Dimension parameters of OR-SRM.

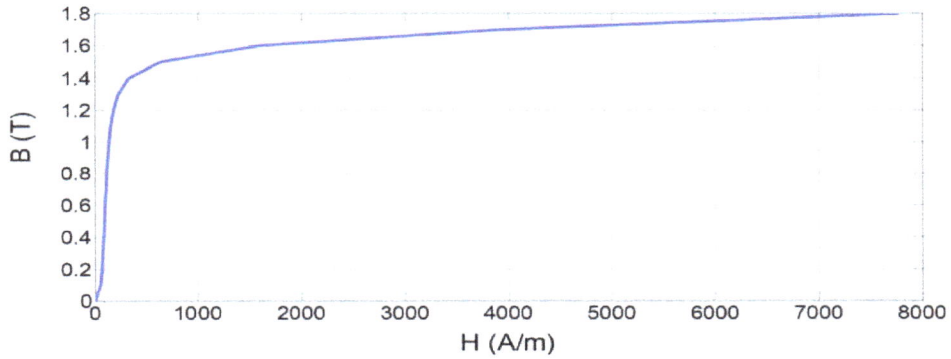

Figure 3. B-H characteristic of M470-50A silicone sheet.

The flux density of stator pole (B_s) is assumed as equal to B_{max}. Omitting leakage flux and packaging factor, stator pole area (A_s) is estimated as

$$A_s = \left(\frac{D}{2} - g\right) \cdot L \cdot \beta_s \tag{2}$$

Flux in stator pole is

$$\varnothing = B_s \cdot A_s \tag{3}$$

Flux in yoke is

$$\varnothing_y = \frac{\varnothing}{2} = \frac{B_s \cdot A_s}{2} \tag{4}$$

Area of yoke is

$$A_y = C_{in} \cdot L \tag{5}$$

Flux density of yoke is

$$B_y = \frac{Q_y}{A_y} \tag{6}$$

Length of stator pole is

$$h_s = \left(\frac{D}{2} - g - \frac{D_{sh}}{2}\right) - \frac{A_y}{L} \tag{7}$$

Length of rotor pole is

$$h_r = \frac{D_0}{2} - C - \frac{D}{2} \tag{8}$$

Magnetic field intensity in air gap is calculated by

$$H_g = \frac{B_g}{4 \cdot \pi \cdot 10^{-7}} \tag{9}$$

B-H curve is determined as a function; then, magnetic field density values corresponding to B_s, B_g, B_r, and B_{rc} are obtained by this function. Consequently, the flux path equations are estimated. Thus, path length of rotor region is

$$l_r = h_r + \frac{C}{2} \tag{10}$$

Path length of rotor yoke is

$$l_{ry} = (2 \cdot \pi \cdot NP \cdot 0.5 \cdot (D_0 - C))/N_s \tag{11}$$

where NP refers to the number of phase.

Path length of air gap region is

$$l_g = g \tag{12}$$

Path length of stator region is

$$l_s = (0.25 \cdot D - 0.5 \cdot g + 0.5 \cdot h_s - 0.25 \cdot D_{sh}) \tag{13}$$

Path length of stator core is

$$l_{sy} = (0.5 \cdot (D_{sh} + C_{in}) \cdot 2 \cdot \pi \cdot NP)/N_s \tag{14}$$

In this situation, total ampere-turns can be calculated as

$$TAT = 2 \cdot (H_s \cdot l_s + H_r \cdot l_r) + \left(B_g \cdot \frac{A_g}{P_a} \right) + \left(H_{ry} \cdot l_{ry} + H_{sy} \cdot l_{sy} \right) \tag{15}$$

where P_a refers to permeability.

Turn per phase is

$$T_{ph} = \frac{TAT}{I_p} \tag{16}$$

where I_p refers to rated current.

Current rate is

$$i_p = \frac{TAT}{T_{ph}} \tag{17}$$

Accurate estimation of inductances, L_a and L_u, is crucial for reliable design. High estimation error of inductances causes the torque to be calculated wrong. Therefore, to calculate the inductances (L_a and L_u) at aligned and unaligned positions, respectively, 2 base and 7 base flux paths are selected. The rest of the fluxes can cause rising of the estimation complexity, and variation of the result will be trivial, as shown in **Figures 4** and **5**.

Tube 1 and Tube 7 flux paths are employed to calculate L_a. Eqs. (13) and (14) calculate ampere-turns of Tube 1 and Tube 7, namely. Eq. (15) estimates the inductance at aligned position. Eq. (17) refers to calculation of leakage inductance in Tube 7. Herein, B_{smin} is the value equalizing Eq. (14) to Eq. (16). Consequently, Equation (18) calculates the inductance of a single pole pair at aligned position:

$$TAT_1 = 2 \cdot (H_s \cdot l_s + H_r \cdot l_r) + \left(B_g \cdot \frac{A_g}{P_a} \right) + \left(H_{ry} \cdot l_{ry} + H_{sy} \cdot l_{sy} \right) \tag{18}$$

$$TAT\_7 = H_s \cdot l_s + \frac{B_s \cdot A_{sf}}{P_f} + H_{sy} \cdot l_{sy} \tag{19}$$

$$L_1 = TAT\_1 \cdot \frac{B_s \cdot A_s}{i_p^2} \tag{20}$$

$$TAT_7 = \left(\frac{3}{4} \right) \cdot \frac{1}{2} \cdot T_{ph} \cdot i \tag{21}$$

$$L_7 = TAT\_7 \cdot \frac{B_{smin} \cdot A_{sf}}{i_p^2} \tag{22}$$

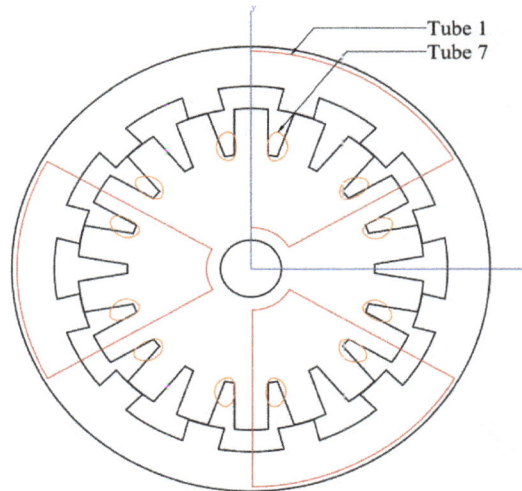

Figure 4. Flux paths for inductance calculation of OR-SRM at aligned position (L_a).

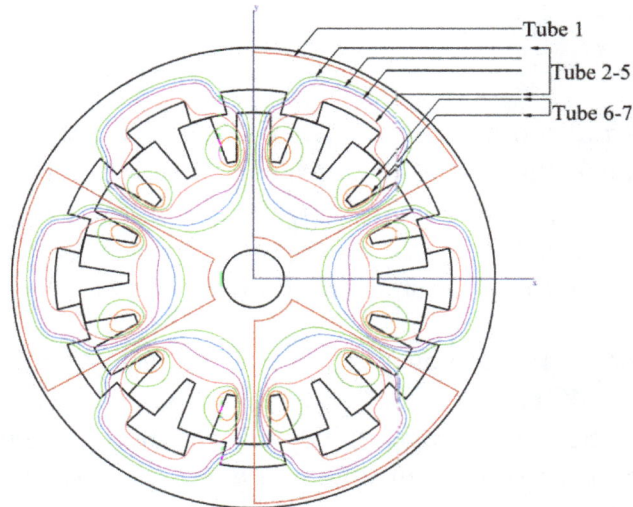

Figure 5. Flux paths for inductance calculation of OR-SRM at unaligned position (L_u).

$$L_a = L_1 + 4 \cdot L_7 \tag{23}$$

At this stage, the aligned inductance for a single phase is obtained by

$$L_{a\_phase} = N_{tpp} \cdot 0.5 \cdot L_a \tag{24}$$

where N_{tpp} represents the number of teeth per phase.

The flux paths in **Figure 5** estimate L_u for a single pole pair. Equations (20)–(23) calculate total ampere-turns of each flux path. Substituting B_{smin} into Eq. (24), inductance of its own flux path is obtained:

$$TAT_1 = T_{ph} \cdot i = 2 \cdot (H_s \cdot l_s) + \frac{B_s \cdot A_{1s}}{P_1} + \frac{H_{sy} \cdot l_{sy}}{2} + \frac{H_g \cdot l_g}{2} \tag{25}$$

$$TAT_n = 2 \cdot (H_{sn} \cdot l_{sn} + H_{rn} \cdot l_{rn}) + \frac{B_{sn} \cdot A_{sn}}{P_n} + \ldots (H_{ryn} \cdot l_{ryn} + H_{syn} \cdot l_{syn}) (n = 2, 3, 4, 5) \tag{26}$$

$$TAT_6 = \left(\frac{1}{2}\right) \cdot \frac{5 \cdot T_{ph} \cdot i}{4} = 2 \cdot H_s \cdot l_s + \frac{B_s \cdot A_{6s}}{P_6} + H_{sy} \cdot l_{sy} \tag{27}$$

$$TAT_7 = \frac{T_{ph} \cdot i}{4} = H_s \cdot l_s + \frac{B_s \cdot A_{7s}}{P_7} + H_{sy} \cdot l_{sy} \tag{28}$$

$$L_{u(k)} = TAT_{(k)} \cdot \frac{B_{smin(k)} \cdot A_{(k)s}}{i_p^2} \ (k = 1, 2, 3, 4, 5, 6, 7) \tag{29}$$

Finally, inductance of a single pole pair at unaligned position is calculated by

$$L_u = L_{u1} + 2 \cdot (L_{u2} + L_{u3} + L_{u4} + L_{u5}) + 4 \cdot (L_{u6} + L_{u7}) \tag{30}$$

At this stage, the unaligned inductance for a single phase is obtained by

$$L_{u\_phase} = N_{tpp} \cdot 0.5 \cdot L_u \tag{31}$$

Energy calculation at aligned and unaligned position, average torque of pole pair, and average torque are all, respectively, given as

$$W_a = \frac{1}{2} \cdot L_a \cdot i_p^2 \tag{32}$$

$$W_u = \frac{1}{2} \cdot L_u \cdot i_p^2 \tag{33}$$

$$T_{pp} = \frac{((W_a - W_u) \cdot N_s \cdot N_r)}{N_{tpp} \cdot \pi} \tag{34}$$

$$T_{average} = T_{pp} \cdot N_{tpp} \cdot 0.5 \tag{35}$$

The coil length of pole pair is obtained by

$$l_{pp} = 4 \cdot \left[L + \left(\frac{(0.5 \cdot D - g - 0.5 \cdot h_s) \cdot 2 \cdot \pi}{18} \right) \right] \cdot T_{ph} \tag{36}$$

Conductor area of conductor diameter (0.95) is given as

$$CA = (0.95)^2 \cdot 0.25 \cdot \pi \cdot NR \tag{37}$$

where NR refers to the number of wires.

Pole and phase resistances are, respectively, estimated by

$$R_{s1} = (0.0177 \cdot I_{pp})/CA \tag{38}$$

$$R_s = N_{tpp} \cdot 0.5 \cdot R_{s1} \tag{39}$$

Copper loss is calculated in

$$P_{cl} = (I_p)^2 \cdot R_s \tag{40}$$

And, iron volume is provided as

$$D_{vol} = \left[\left((A_s \cdot h_s \cdot N_s) + (A_r \cdot h_r \cdot N_r) \right) + \pi \cdot L \cdot \left((0.5 \cdot D - g - h_s)^2 + (0.5 \cdot D_0)^2 - (0.5 \cdot D_0 - C)^2 \right) \right] - V_{cool} \tag{41}$$

where V_{cool} is the total volume of the cooling holes.

Torque/volume rate is given in

$$TVolume = T_{average}/D_{vol} \tag{42}$$

Efficiency is obtained by

$$\eta(\%) = ((P_i - P_{cl})/P_i) \cdot 100 \tag{43}$$

2.3. Optimization of motor dimension parameters using genetic algorithm

Calculation of dimension parameters for SRM could be optimized by trial and error, but estimated cost of these types of methods is fairly high. And also, convergence could not be manually accomplished in every experiment including specified trials. Because tiny tuning inputs on the dimension parameters can emerge large variations on the motor torque and its efficiency.

Exploring optimum values of five different parameters, $f_{obj}(\beta_s, \beta_r, D, C_{in}, C)$, to be involved in the object functions is NP-complete problem. Therefore, evolutionary algorithm is proposed to solve NP-complete combinatorial optimization problems. Genetic approach generates better individuals influencing torque and efficiency for each new generation. Stages of the proposed GA are given by:

2.3.1. Coding individuals

Thirty-two-bit binary codes for each individual are utilized to build genes. Each of parameters, $\beta_s, \beta_r, D, C_{in}$, and C are represented by five bits, all of them stands for chromosome.

The restrictions of the optimized parameters are taken into consideration as below [33]:

- Rotor pole arc is selected to be equal or greater than stator pole arc, since the number of rotor pole is less than the number of stator pole.

- βs should be equal or greater than step angle to generate required torque using OR-SRM. When βs is selected as smaller than step angle, none of the phases may not have rising inductance slope.

- Rotor pole angle should be greater than the sum of stator and rotor pole arc.

- Other parameters such as magnetic saturation and physical strength are also considered in these restrictions.

We have used 25 bits out of 32 bits as shown in **Figure 6**. Out of range values are not included in the solution space. The constraints with regard to motor dimension parameters are also involved in fitness function to reduce solution space, which are given as.

$$\beta_s + \beta_r < \left(\frac{360}{N_r}\right), \beta_s \leq \beta_r, \text{torque} > 400$$

$$80 < \text{efficiency} < 100, 0.4 < \beta_r - \beta_s < 1.8$$

$$11 \leq \beta_s \leq 15, 12 \leq \beta_r \leq 16, 0.160 \leq D \leq 0.185$$

$$0.012 \leq C \leq 0.032, 0.06 \leq C_{in} \leq 0.091, 0 \leq P \leq \left(2^{25} - 1\right)$$

Considering constraints above, five bits of D, C, C_{in}, β_r, and β_s are sufficient to get a result in required precision.

2.3.2. Initial population

The population consists of n individuals each of which includes aforementioned five parameters. n is taken as 200. Allowing the entire range of possible solutions, they are randomly produced by

$$P = \text{int}\left(\text{rand}(\ )\cdot\left(2^{25} - 1\right)\right) \tag{44}$$

2.3.3. Generating offspring

The fitness function is defined by considering required torque and efficiency that are, respectively, 450–500 Nm and over 80%.

Estimation of fitness rate: Fitness function is designated by required torque and efficiency of motor as

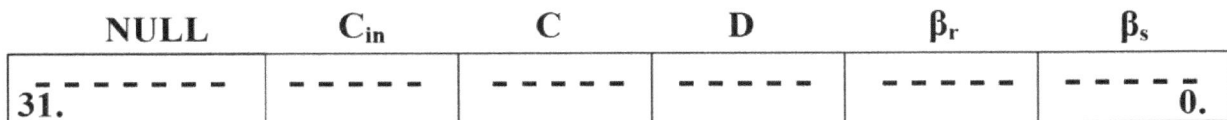

NULL	C_{in}	C	D	β_r	β_s
31. -----	-----	-----	-----	-----	----- 0.

Figure 6. Chromosome of each individual.

$$fitness = \left(\left(\frac{Torque}{400} \right) + \left(\frac{efficiency}{100} \right) \right) \cdot 100 \qquad (45)$$

The individuals are sorted in fitness values. Then, the individuals, whose the fitness values are better, remain in the population, and the ones whose fitness values are worse are eliminated. Therefore, in each iteration, quarter of the population turns out to be eliminated. Sorted population is shown in **Figure 7**.

Crossover operator: Random 2 bits out of least significant 25 bits of parents are crossed to generate a new offspring couple. The chromosome of parents and offspring are given in **Figure 8(a)** and **(b)**.

Mutation operators: Mutation rate is specified as 0.05 to reach global solution.

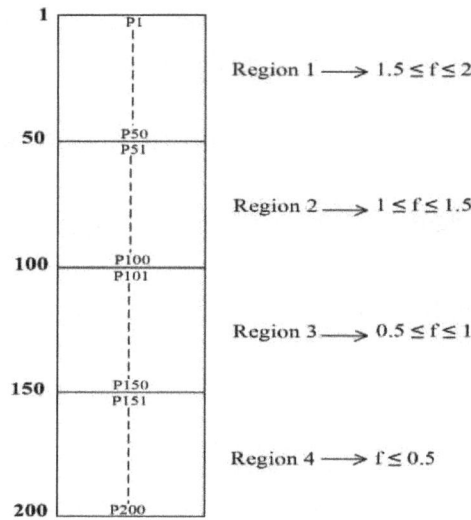

Figure 7. Sorted population.

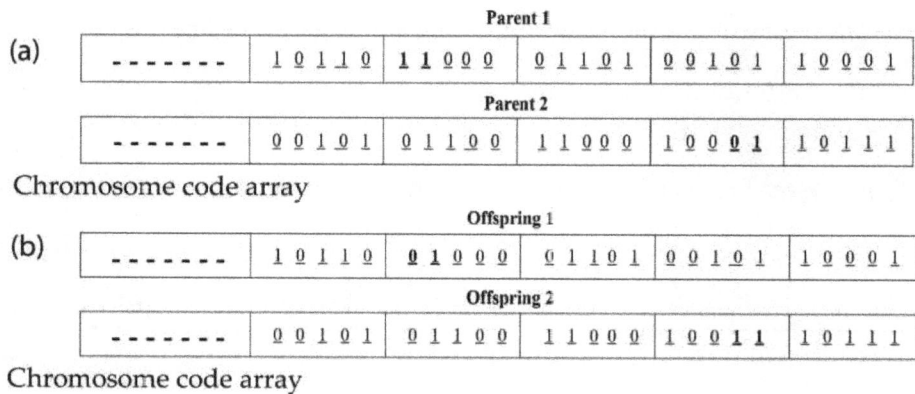

Figure 8. (a) Chromosome of parents, (b) Chromosome of offspring.

Optimized parameters								Electrical parameters			Fitness	
β_r (deg.)	β_s (deg.)	D (mm)	C (mm)	C_{in} (mm)	h_s (mm)	h_r (mm)	D_{sh} (mm)	R_f (Ω)	L_u (mH)	L_a (mH)	T (Nm)	η (%)
12.26	11.13	182.7	21.7	76	66.2	20.6	80	0.33	6.4	20.7	467.72	93.69

Table 2. Estimated parameters for the manufactured OR-SRM.

2.3.4. Termination criteria

The derivation of the fitness values is calculated by

$$\varepsilon = \sum_{i=1}^{n/2} \left(f_i - f_{i-1} \right) \tag{46}$$

Here, termination criteria are determined as $\varepsilon \leq 0.001$.

Considering fitness constraints, optimized values of five parameters, correspondent motor dimension, and electrical parameters are given in **Table 2**.

3. The analysis of motor using Maxwell 3D package software

To validate OR-SRM design parameters, Maxwell 3D analysis is carried out following manufacturing stage. Field distribution, inductance, and torque curves are consecutively plotted by Maxwell 3D software as shown in **Figures 9–11**. MATLAB analytical solution results are verified with Maxwell 3D results as given in **Table 3**.

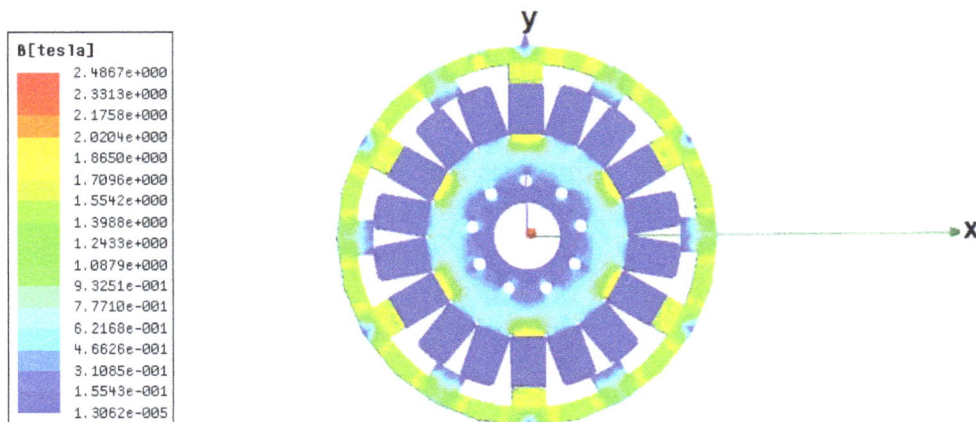

Figure 9. Field distribution of 18/12 OR-SRM.

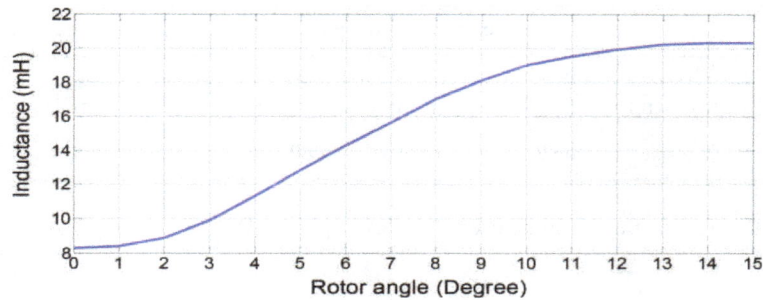

Figure 10. Inductance curve of 18/12 OR-SRM.

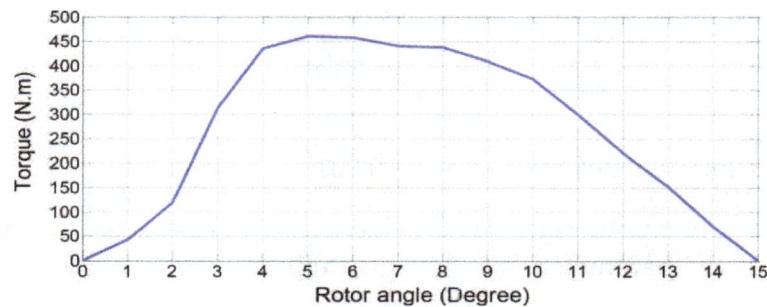

Figure 11. Torque curve of 18/12 OR-SRM.

	GA Solution	Maxwell 3D	Error %
Torque (Nm)	467.72	460.47	1.57
Efficiency (%)	93.69	95.28	1.6

Table 3. Comparison of GA solution and Maxwell 3D results.

4. Experimental results and discussions

OR-SRM has been manufactured by using motor sizes optimized via MOGA. The motor at manufacturing stage is given in **Figure 12**. OR-SRM is cooled via cooling channels shown in **Figure 12** as the forced circulation air cooling. A nematic air dryer compressor is used to cool the OR-SRM. Thus, the temperature of the motor is prevented by these cooling channels to reach too high levels. Therefore, the changes in the winding resistance are minimized. Its negative effect on the efficiency is also reduced.

The manufactured OR-SRM is shown in **Figure 13**.

In this study, asymmetric half-bridge converter is designed and used to feed OR-SRM. A capacitor charged and discharged by switching IGBT is connected between the battery and

Figure 12. Stator windings of OR-SRM at manufacturing stage.

Figure 13. Manufactured OR-SRM.

converter. According to different speeds and reference currents, trigger angles of each phase winding are calculated. The trigger angles are correlated to the speed, battery voltage, reference current, and rotor position.

To compare the estimated torque obtained by Maxwell 3D and that of the manufactured motor, locked rotor experiment has been carried out. Herein, torque rate is measured by torque sensor mounted on tested while streaming the coil current, 75 A by locking OR-SRM rotor at designated positions.

Test bed of OR-SRM is shown in **Figure 14**. OR-SRM is loaded with an induction motor controlled by the direct torque control driver. Torque sensor is used for measuring torque of OR-SRM. The features of torque sensor have very short construction, broad input voltage range, current, and output voltage; the measurement accuracy is less than 0.5% of full scale; and measurement ranges change from 5 to 500 Nm. The control algorithm is run by dSPACE DS1103. Hall-effect current and voltage sensors are used for measuring motor phase and battery currents and battery voltage.

The comparison of Maxwell 3D and experimental results have been drawn in **Figure 15**.

Average rate of relative errors between Maxwell 3D and experimental results is estimated as 2.248%, for the angles at designated positions.

Figure 14. OR-SRM test bed.

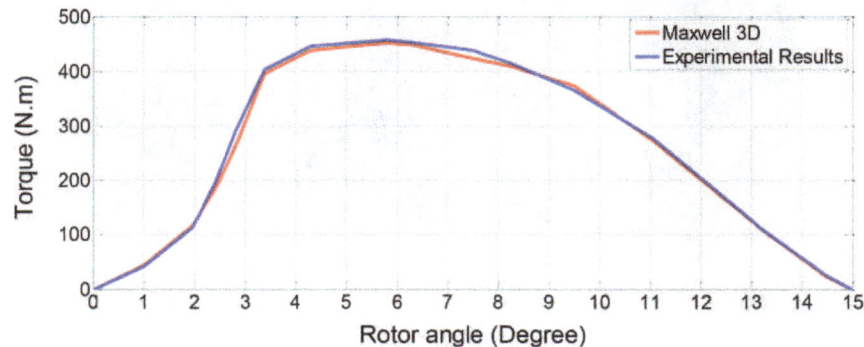

Figure 15. Comparison of Maxwell 3D and experimental results.

In this paper, we have proposed an optimum solution satisfying the torque and efficiency of OR-SRM via MOGA considering constrains of EV dynamics. The efficiency and torque values are the components of objective function. Torque density has not been involved in the objective function since the package length and outer diameter are specified to fit into 21-inch wheel rim. The solution of torque ripple problem might be considered by varying trigger angles in motor control stage rather involving it during OR-SRM design stage, which is well-known motor control method. As a future work, torque ripple will be involved in multi-objective optimization function via analysis of dynamic solution.

5. Conclusion

In this study, a direct drive OR-SRM mounted in wheel is designed and manufactured for the subject EV. The dimensional and electrical parameters of 18/12 OR-SRM for EV have been calculated by multi-objective evolutionary algorithm. The mathematical equations derived by geometry of OR-SRM have been used to get results from objective functions resided in fitness function. The vehicle dynamics, five independent variables of motor dimension, and the constraints including outer diameter and package length of the motor are employed by evolution algorithm. The multiparameters including stator and rotor pole arc angles, their yoke lengths, and inner diameter of rotor are optimized using objective functions such that the highest efficiency for deserved torque is obtained. Consequently, we can obviously suggest that this approach could be conducted to design a standard-type OR-SRM.

Acknowledgements

The authors gratefully acknowledge the support of the Scientific and Technological Research Council of Turkey (No: 113 M090).

Author details

Zeki Omaç[1], Mehmet Polat[2], Mustafa Kaya[3], Eyyüp Öksüztepe[4], Haluk Eren[4*],
Merve Yıldırım[5] and Hasan Kürüm[5]

*Address all correspondence to: he.edu.tr@gmail.com

1 Electrical and Electronics Engineering, University of Munzur, Tunceli, Turkey

2 Mechatronics Engineering, University of Fırat, Elazığ, Turkey

3 Digital Forensics Engineering, University of Fırat, Elazığ, Turkey

4 School of Aviation, University of Fırat, Elazığ, Turkey

5 Electrical and Electronics Engineering, University of Fırat, Elazığ, Turkey

References

[1] Yildirim M, Polat M, Kürüm H. A survey on comparison of electric motor types and drives used for electric vehicles. In: IEEE 16th International Power Electronics and Motion Control Conference and Exposition (PEMC); 2014. pp. 218-223

[2] Trancho E, Ibarra E, Arias A, Kortabarria I, Prieto P. A practical approach to HFI based sensorless control of PM-assisted synchronous reluctance machines applied to EVs and HEVs. In: IEEE 43rd Annual Conference of the Industrial Electronics Society (IECON); 2017. pp. 1735-1740

[3] Tarek MTB, Choi S. Design and rotor shape modification of a multiphase high speed permanent magnet assisted synchronous reluctance motor for stress reduction. In: IEEE Energy Conversion Congress and Exposition (ECCE); 2007. pp. 5389-5395

[4] Asgar M, Afjei E, Behbahani A, Siadatan A. A 12/8 double-stator switched reluctance motor for washing machine application. In: IEEE 6th Power Electronics, Drives Systems & Technologies Conference (PEDSTC); 2015. pp. 168-172

[5] Tursini M, Villani M, Fabri G, Di Leonardo L. A switched-reluctance motor for aerospace application: Design, analysis and results. Electric Power Systems Research. 2017; **142**:74-83

[6] Öksüztepe E. In-wheel switched reluctance motor design for electric vehicles by using a pareto-based multiobjective differential evolution algorithm. IEEE Transactions on Vehicular Technology. 2017;**66**(6):4706-4715

[7] Santiago JD, Bernhoff H, Ekergard B, Eriksson S, Ferhatovic S, Waters R, Leijon M. Electrical motor drivelines in commercial all-electric vehicles: A review. IEEE Transactions on Vehicular Technology. 2012;**61**(2):475-484

[8] Xue XD, Cheng KWE, Cheung NC. Selection of electric motor drivers for electrical vehicles. In: IEEE Power Engineering Conference, AUPEC'08; Australasian Universities; 2008. pp. 1-6

[9] Wenping C, Mecrow BC, Atkinson GJ, Bennett JW, Atkinson DJ. Overview of electric motor technologies used for more electric aircraft (MEA). IEEE Transactions on Industrial Electronics. 2012;**59**(9):3523-3531

[10] Rahman KM, Fahimi B, Suresh G, Rajarathnam AV, Ehsani M. Advantages of switched reluctance motor applications to EV and HEV: Design and control issues. IEEE Transactions on Industry Applications. 2000;**36**(1):111-121

[11] Ramu K. Switched Reluctance Motor Drives: Modeling, Simulation, Analysis, Design, and Applications. U.S.: CRC Press, Taylor and Francis Group; Dec, 2017

[12] Miller TJE. Electronic Control of Switched Reluctance Machines. U.S.: Elsevier, Newnes; 2001

[13] Yildirim M, Polat M, Oksuztepe E, Omac Z, Yakut O, Eren H, Kaya M, Kurum H. Designing in-wheel switched reluctance motor for electric vehicles. In: IEEE 16th International Power Electronics and Motion Control Conference and Exposition (PEMC); 2014. pp. 793-798

[14] Sengor I, Polat A, Ergene LT. Design and analysis of switched reluctance motors. In: IEEE 8th International Conference on Electrical and Electronics Engineering (ELECO); Nov. 2013. pp. 586-590

[15] Sakthivel P, Chandrasekar V, Arumugam R. Design of a 250w, low speed switched reluctance hub motor for two wheelers. In: IEEE 1st International Conference on Electrical Energy Systems (ICEES); Jan. 2011. pp. 176-181

[16] Koibuchi K, Ohno T, Sawa K. A basic study for optimal design of switched reluctance motor by finite element method. IEEE Transactions on Magnetics. 1997;**33**(2):2077-2080

[17] Jawad F, Shahgholian G, Ghazizadeh H. Analysis of dynamic behavior of switched reluctance motor-design parameters effects. In: IEEE 15th Mediterranean Electrotechnical Conference; 2010. pp. 532-537

[18] Takano Y, Takeno M, Hoshi N, Chiba A, Takemoto M, Ogasawara S, Rahman MA. Design and analysis of a switched reluctance motor for next generation hybrid vehicle without PM materials. In: IEEE International Power Electronics Conference (IPEC); June 2010. pp. 1801-1806

[19] Chiba A, Takano Y, Takeno M, Imakawa T, Hoshi N, Takemoto M, Ogasawara S. Torque density and efficiency improvements of a switched reluctance motor without rare-earth material for hybrid vehicles. IEEE Transactions on Industry Applications. 2011;**47**(3): 1240-1246

[20] Mirzaeian B, Moallem M, Tahani V, Lucas C. Multiobjective optimization method based on a genetic algorithm for switched reluctance motor design. IEEE Transactions on Magnetics. 2002;**38**(3):1524-1527

[21] Yoshiaki K, Kosaka T, Matsui N. Optimum design approach for a two-phase switched reluctance compressor drive. IEEE Transactions on Industry Applications. 2010;**46**(3): 955-964

[22] Omekanda AM. Robust torque and torque-per-inertia optimization of a switched reluctance motor using the Taguchi methods. IEEE Transactions on Industry Applications. 2006;**42**(2):473-478

[23] Cosovic M, Smaka S, Salihbegovic I, Masic S. Design optimization of 8/14 switched reluctance machine for electric vehicle. In: IEEE XXth International Conference on Electrical Machines (ICEM); 2012. pp. 2654-2659

[24] Naayagi MR, Kamaraj V. Optimal design of switched reluctance machine using genetic algorithm. Optimization. 2002;**19**(1):350-353

[25] Behzad MD, Moallem P. Genetic algorithm based optimal design of switching circuit parameters for a switched reluctance motor drive. In: IEEE International Conference on Power Electronics, Drives and Energy Systems (PEDES); 2006

[26] Costa MC, Nabeta SI, Dietrich AB, Cardoso JR, Marechal Y, Coulomb JL. Optimisation of a switched reluctance motor using experimental design method and diffuse elements response surface. IEE Proceedings - Science, Measurement and Technology. 2004;**151**(6): 411-413

[27] Zhang Y, Xia B, Xie D, Koh CS. Optimum design of switched reluctance motor to minimize torque ripple using ordinary Kriging model and genetic algorithm. In: IEEE International Conference onElectrical Machines and Systems (ICEMS); Aug. 2011. pp. 1-4

[28] Ma C, Qu L. Multiobjective optimization of switched reluctance motors based on design of experiments and Particle Swarm Optimization. In: IEEE Transactions on Energy Conversion. 2015;(**30**)3:1144-1153

[29] Labak A, Narayan CK. Outer rotor switched reluctance motor design for in-wheel drive of electric bus applications. In: IEEE XXth International Conference on Electrical Machines (ICEM); 2012

[30] Xue XD, Cheng KWE, Ng TW, Cheung NC. Multi-objective optimization design of in-wheel switched reluctance motors in electric vehicles. IEEE Transactions on Industrial Electronics. 2010;**57**(9):2980-2987

[31] Hennen MD, Rik WDD. Comparison of outer-and inner-rotor switched reluctance machines. In: IEEE 7th International Conference on Power Electronics and Drive Systems (PEDS); 2007

[32] Kiyota K, Akira C. Design of switched reluctance motor competitive to 60-kW IPMSM in third-generation hybrid electric vehicle. IEEE Transactions on Industry Applications. 2012;**48**(6):2303-2309

[33] Omaç Z, Polat M, Öksüztepe E, Yıldırım M, Yakut O, Eren H, Kaya M, Kürüm H. Design, analysis, and control of in-wheel switched reluctance motor for electric vehicles. Electrical Engineering

3

Lightweight High-Efficiency Power Train Propulsion with Axial-flux Machines for Electric or Hybrid Vehicles

Sorin Ioan Deaconu, Vasile Horga, Marcel Topor, Fabrizio Marignetti, Lucian Nicolae Tutelea and Ilie Nuca

Abstract

The aim of this chapter is to present a new type of powertrain with dimensions and low weight, for vehicles with reduced carbon emissions, which have an axial synchronous machine with one stator and two rotor, with static converter that is simple and inexpensive, a broadcast transmission system using an electric differential, with the control of the two rotors so that they can operate as motor/generator, at the same rotational direction or in opposite directions, at the same speed value, at slightly different speeds or at much different speeds by using an original dual vector control with operating on dual frequency. This is a major concern of hybrid and electric vehicle manufacturers. Expected results: a lighter power train with 20% and an increase in 5% of electric drive efficiency, low inertia rotor at high speed, a compact electric drive system with high torque and simple control, intelligent energy management system with a new vision of technological and innovation development, and equal importance of environment protection. The electrical machines for hybrid (HEV) or electric (EV) drives include a variety of different topologies. According to outcomes of literature survey, induction machines alongside synchronous machines take the major place in HEV or EV power trains.

Keywords: axial-flux machines, dual rotor, single stator, single inverter control, hybrid vehicle, electric vehicle, lightweight power train

1. Introduction

Automobiles use onboard fuels (energy carriers) in order to transport goods and people. The conversion of onboard energy to propulsion energy is performed by the power train. Some parts of this energy may be stored as conservative energy (kinetic or potential energy) in vehicle. Unfortunately, all of these conversion processes cause substantial energy losses and hence high fuel consumption. Nowadays, in order to obtain propulsion energy, most of the vehicles are based on the combustion of hydrocarbon fuels. Theoretically, the complete combustion of chemical fuel generates only heat, which is converted into mechanical energy, and carbon dioxide (CO_2) and water (H_2O), which are released into the atmosphere. These combustion products do not harm the environment [1]. However, actually combustion of hydrocarbon fuel is never complete, resulting also in a certain amount of nitrogen oxides (NO_x) and unburned hydrocarbons, all of these impacting people's health and the environment. Furthermore, even if CO_2 is assimilated by plants and captured by sees and oceans, these natural assimilation processes are saturated, which cause an accumulation of it in the atmosphere. These gases block re-reflected infrared radiation of the Earth, which comes from the Sun, and in this way keep the energy in the atmosphere (greenhouse effect). This energy increases the global temperature and causes climate change. Therefore, over the past decades, research and development activities related to road transport have highlighted the need to develop less polluting and safer transport. Because pollutant emissions and fuel consumption are directly proportional, a cleaner vehicle means a fuel-efficient vehicle. Different methods and tools, which analyze and assess the use of resources and impacts on the environment, are available. These are referred to as environmental system analysis tools. They can be categorized according to their object in focus (policies, plans, products, and functions or substances) and their studied impacts (natural resources and/or environmental impacts) [2]. Energy analysis, one of the methods suited to analyze the use of natural resources, is focused on energy or material flows (with a focus on input flows) in energy or physical units. Another very useful method is represented by the Life Cycle Assessment (LCA) which is based on an environmental assessment and is able to evaluate the impact of materials/products on the environments. LCA usually operates on the energy usage, environmental emissions (kg CO_2/kg of material), and the amount of materials used to make the final product. An LCA analyzes the potential environmental impacts of a product or a service along its entire life cycle. The life cycle includes all of phases (from cradle to grave): the raw material extraction, the production, use and any end-of-life treatment including recycling (ISO 14040:2006). For a passenger car, analyzed in use phase, a comprehensive analysis of the energy consumption involves at least three energy conversion steps: well-to-tank (WTT), tank-to-vehicle (TTV), and vehicle-to-miles (VTM) [3]. In a first step, WTT, the primary energy carriers (chemical energy in fossil hydrocarbons, solar radiation used to produce biomass or electric energy, nuclear energy, etc.) are converted into an energy carrier that is suitable for onboard storage, that is, to a "fuel" (examples are gasoline, hydrogen, etc.). Then, in the second step—TTV—this "fuel" is converted by the propulsion system to mechanical energy. The third energy transformation—VTM—is determined by the vehicle parameters and the driving profile. In this step, the mechanical energy produced in the second conversion step is ultimately dissipated to thermal energy that is deposited to the ambient. So, there are essentially three possible approaches to reducing the total energy consumption of passenger cars: improve the WTT efficiency, the TTV efficiency, and the VTM efficiency. *WTT efficiency improvement:* deriving from crude oil, the gasoline and diesel are the major liquid fuels for internal combustion engine vehicles (ICEVs). Electric vehicles (EVs) are an excellent solution to rectify this unhealthy dependence because electricity can be generated by almost all kinds of energy resources: thermal power (oil, natural gas, and coal), nuclear power, hydro power, wind power, solar power, oceanic power, geothermal power, and biomass power. Even taking into account the emissions from refineries to produce gasoline for ICEVs and the emissions from power plants (PPs) to generate electricity for EVs, the overall harmful emissions of EVs are still much lower than those of ICEVs. The reduction of CO_2 emission can reach a level about 5% with the adoption of EVs and high-efficient PPs. The level of efficiency increase may be further extended when higher percentages of clean or renewable power generation are used. Certainly, the increase may even be negative when inefficient coal-fired PPs are used [4, 5]. *TTV efficiency improvement:* several possible directions are available for improvement: improvement on the component level and improvement on the system level [3], so we can mention here some of them: one may consider an improvement on the peak efficiency of the power train components; also, there is a possibility to improve the part-load efficiency of the power train components, and similarly, one may add the capability to recuperate the kinetic and potential energy in order to store it in the vehicle; also, there is another possible direction that considers the need to optimize the structure and the parameters of the propulsion system. All of this presumes the fact that the fuel(s) used and the vehicle parameters are fixed, and the control systems realize appropriate supervisory control algorithms that take advantage of the opportunities offered by the chosen propulsion system configuration. By taking crude oil as 100%, the total energy efficiencies (well-to-vehicle) for EVs and ICEVs are 18 and 13%, respectively. Therefore, even when all electricity are generated by oil-fuelled PPs, EVs are more energy efficient than ICEVs by about 40% [5]. Moreover, EVs possess a definite advantage over ICEVs in energy utilization, namely regenerative braking; EVs can recover the kinetic energy during braking and utilize it for battery recharging, whereas ICEVs wastefully dissipate this kinetic energy as heat in the brake discs and drums. With this technology, the energy efficiency of EVs is virtually boosted up by further 10% [4]. *VTM efficiency improvement:* there are a few ways to reduce fuel consumption in the vehicle, such as improved power train efficiency, new and advanced power trains like internal combustion-diesel hybrids and fuel cell vehicles, adoption of alternative fuels, but VTM efficiency improvement can be made by improving aerodynamics, and lightweight design. The weight of vehicles has increased continuously in the past four decades. Drivers for the weight increase are higher demands—of customers or legislation—on safety, performance, comfort, reliability, and other vehicle characteristics. These demands lead to additional and more complex parts in each new vehicle generation. Anyway, a spiral effect that sums all the recent technological outcomes forces an increase of the weight at a higher level even higher. Due to the interlinks and dependencies that are presented between components, it is possible to have a weigh increase. Weight increase is a tricky problem since it leads to the requirement for a more powerful and heavier power train and electric motor or engine. The heavier load on the chassis and the dynamic demands on the power train impose the need of reinforcements and an extra weight increase to the car manufacturers. In addition, in order to maintain the driving range, extra energy storage is needed. From the mechanical point of view, the stiffness of the vehicle body must be reconsidered again which again impose a supplementation in the available power of the engine, which is equal to the fact that we need a more powerful motor or an engine. And since the power density cannot be modified, we have a spiral effect called secondary weight effect. The good news is that if we want to reverse the case, general weight reductions turn the spiral effect around and lead to secondary weight reductions [2]. Generally speaking, the major automakers have adopted advanced technologies capable to improve the fuel economy of their vehicles, and among this, the reduction of the weight is still one of the approaches they make use very often. This is not only because this method improves fuel economy but also because of the emergence of advanced materials that may result in lower costs in association with good manufacturability. Weight reduction or lightweighting can be reached through careful redesign, component downsizing, alternative material substitution, multiple integrating parts and/or functions, or a combination of all these methods [6]. Redesigning aims at reducing aerodynamic drag force and rolling resistance, while downsizing focuses more on reducing the dimensions of the vehicle. Material lightweight design portrays a special case of lightweight design as it does not focus on the mere reduction of material but on the substitution of materials. In this way, a material is substituted with one of lighter densities or with a material of better properties. These properties can be the strength, a smaller distortion, or a reduced wear (e.g., replacement of conventional steel with a high strength steel). Functional lightweight design chooses either a strategy of integrating several parts and/or functions into one component or of separating the functions to achieve a lower weight. Vehicle mass is key factor for a lightweight design as well as in the operational performance of the vehicle in terms of mechanical resistance to the road and especially in the fuel economy. Optimization of the mass of the vehicle not only limits the friction force with roads by reducing rolling resistance but also modifies in a positive way the acceleration resistance and climbing resistance. A general rule of thumb says that a 100-kg savings in vehicle mass will result in a fuel savings of 0.12–0.15 l/100 km and 0.85–1.4 kg CO_2/100 km for ICEVs [7] or 0.347 kWh/100 km/100 kg for EVs [8]. Lightweight solutions reduce the energy demand of the vehicle in the use phase. The mix of both measures—the manufacturing of lightweight electric vehicles (LEVs)—has the potential leads to conduct to an even higher savings and a spectacular reduction of environmental impacts in comparison to the conventional solutions. However, aspects like the integration of different materials and more complex and energy-intensive production processes can also lead to higher environmental impacts. From what is known today, the production of the components for the electric drive train—particularly the battery—is energy-intensive process and uses complex chemical assemblies which demand complex recycling processes. Similarly, lightweight materials are usually more energy-intensive than the conventional material steel and can be less suitable for recycling. Thus, to avoid burden shifting—which means one type of emission is reduced while another is increased—it is necessary to assess all relevant types of emission. The environmental assessment over the entire life cycle—an LCA—calculates these trades-off. One of the promising approaches is the functional lightweight design. Optimized power train systems and appropriate control algorithms are instrumental to achieve this objective.

2. EV lightweight propulsion systems configuration

The EV offers the definite advantages of zero roadside emissions and minimum overall emissions (taking into account the emissions due to electricity generation by PPs). Previously, the EV was mainly converted from the ICEV, simply replacing the internal combustion engine IC (diesel of gas) with an electric motor (comparable in terms of power capability), while all the other components are kept the same. It is noticed early that the converted EV has not been a good solution because of the drawback of heavy weight, loss of flexibility, and degradation of performance/reliability. Currently, the modern EV is a dedicated built system. This EV configuration is specifically designed and is based on the original body and frame configurations capable to satisfy the structural requirements which are unique to EVs. An EV or HEV car needs a body and a frame which has to take advantage of the greater flexibility of electric propulsion [1, 5]. Compared with the ICEV, the configuration of the EV is particularly extremely flexible and thus is more capable to sustain body and frame improvements. This flexibility is the effect of several factors which are unique to the EV. The first characteristic element is represented by the fact that the energy/power flow in the EV is done mainly by flexible electrical wires rather than by mechanical couplings or rigid shafts. Thus, the concept of *distributed subsystems* or decentralization in the EV is really achievable. Secondly, different EV propulsion arrangements (such as independent four-wheel and in-wheel drives) involve a significant difference in the system configuration/organization. Thirdly, different EV energy sources have different weights, sizes, and shapes. The corresponding refueling systems also involve different hardware and mechanism. For EV propulsion, the system configuration can be a single motor or a multiple motor. The single-motor configuration uses only one electric motor and one PWM inverter, which can minimize the corresponding size, weight, and cost. For single-motor configuration, the first alternative consists of an electric motor, a clutch, a multi-speed gearbox, and a differential. By incorporating both clutch and multi-speed gearbox, the driver can shift the gear ratios and hence the torque going to the wheels. The differential enables the wheels to be driven at different speeds when cornering—the outer wheel covering a greater distance than the inner wheel. It should be noted that the driveline transmission losses of the transmission gear and mechanical differential can be up to 20% of the total power generated by the motor. With an electric motor that has constant power in a long speed range, a multi-speed gearbox can be replaced by a fixed gearing and reduce the need for a clutch. Because of the absence of a clutch and shifting gears, it can significantly improve the transmission efficiency and reduce the overall size, hence increasing both the energy efficiency and power density. In the third configuration, which is similar to the concept of transverse front-engine front-wheel drive of the existing ICEVs, the electric motor, fixed gearing, and differential are integrated into a single assembly, while both axles point at both driving wheels. The whole drive train is further simplified and compacted. The multiple-motor configuration uses multiple motors to independently drive individual wheels. A dual-motor configuration consists of two electric motors, two PWM inverters, and two optional fixed gears depending on whether using a direct drive or not. Since the two motors are independently controlled, the differential action can be electronically achieved, thus eliminating the bulky and heavy mechanical differential. For modern EVs, the use of in-wheel motor (hub motor) drives is becoming more and more attractive. Adopting a direct drive for in-wheel propulsion not only offers the differential action but also facilitates, many advanced vehicular functions such as the antilock braking system, anti-slip regulation, and electronic stability program. Furthermore, since regenerative braking can take place in each driving wheel, the corresponding energy recovery becomes more effective, which can extend the EV driving range.

3. HEV lightweight propulsion systems configuration

At this moment, EVs possess some noticeable advantages against conventional ICEVs, such as high-energy efficiency and zero environmental pollution. However, when we compare the actual performance, and we are focusing upon the operation range per battery charge, we can observe that the EV is far less competitive than ICEVs. This fact is generated to the lower energy content of the batteries versus the extremely generous energy content of gasoline. Typically, for a passenger car under urban driving with air-conditioning, an EV using present batteries technology (that are heavy and bulky) can travel about 120 km per charge, whereas an ICEV can offer about 500 km per refuel. With such a short driving range per charge, the EV will suffer from the problem of range anxiety. Furthermore, differing from the ICEV, the EV takes time for battery charging. The short-term solutions are hybrid electric vehicles (HEVs). HEVs provide an opportunity for synergism with both technologies in order to have the advantages of both ICEVs and EVs and to overcome their disadvantages. In this category of synergistic exploitation, HEVs use two power sources—a mechanical power source and an electric one. Compared with the EV, the HEV can offer a comparable driving range of the ICEV and use the existing refueling infrastructure of the ICEV, but sacrificing the merits of zero roadside emissions. Given the nature of both mechanical and electrical powers, two classic layouts were most commonly used for the HEVs—hybrid versions of both series and parallel connections. The first one is regarded as the most common and least complicated type of HEVs. Having a direct connection between the engine and the generator makes it easier to fully convert the output mechanical power into electricity via the generator. The resulted electricity is used for either feeding the electrical motor (which is used to put the driving wheels into motion) or keeping the battery charged, all controlled using the load condition. Increased flexibility is the clear advantage in locating the engine-generator set, mainly due to the electrical wiring. Furthermore, the EV propulsion system's flexibility is also included in this configuration. The engine capability can easily be enhanced to a maximum because of its feature that allows it to operate within a precise range of speeds. The main difference between the series hybrid and the parallel hybrid is that the latter permits the driving wheels to be propelled by the parallel power delivered by both electric motor and the engine. The wheels have both the engine and motor connected to their driveline, so the propulsion power that is supplied is a result of the engine on its own, the electric motor on its own, or a combination of both. A smaller engine can operate under conditions yielding higher efficiency. Compared to an EV, for the same performance, the battery can also be downsized. Another advantage of a parallel hybrid is the single energy conversion for both electrical and mechanical. The parallel hybrid can respond to the demand for large, near instantaneous changes in either torque or power. In contrast, the series hybrid is slower. The fast response is an advantage in traffic. The detrimental effects of having two systems are partially mitigated by the fact that many components can be downsized. This way, a parallel hybrid, which can offer all hybrid features (electric launch, idle stop-start, regenerative braking, and downsized engine), enables an improvement of 30% in fuel economy. Series-only or parallel-only designs often do not meet performance requirements. As hybrid technology developed, the utility of series or parallel design became less significant. Mixed designs, rather than series or parallel designs, offer more flexibility [6]. The structure for series–parallel hybrid type blends features from both its components, but with an additional element compared to each of them taken on their own: a supplementary generator compared to the parallel hybrid and another mechanical connection compared to the series hybrid. Even if it holds the beneficial elements for both series and parallel hybrids, the combination between the two is more expensive and more complex. Pure electric and hybrid propulsion are additionally mixed into the complex hybrid. Among the disadvantages of parallel arrangement are the various added power train parts such as added clutches and transmissions that increase the weight. On the other hand, weight is a major impediment to performance. Thus, HEVs, by their very nature, are heavier than ICEVs. For HEVs, performance is measured by mileage and distance covered in electric-only mode. Another important factor is reduced emissions. The introduction of a gasoline HEV reduces CO_2 emissions by 27% for a lightweight car (900 kg) and 20% for a heavy vehicle (2500 kg). On the other hand, the introduction of a diesel HEV reduces CO_2 emissions by 24% for a lightweight car and 18% for a heavy vehicle. HEVs can help reduce CO_2 in a greenhouse gas but are not a panacea [10]. Typically for a HEV, a three-shaft transmission is needed: two input shafts and one output shaft (a three-way gearbox). Because of the need to mechanically connect the engine with the drive shaft, choices for the location of the engine are limited. For the ICEV, the engine rpm and torque can be determined by starting at the tire patch on the road and working upstream to the motor through the power train. In contrast to an ICEV, the HEV needs complex energy management system (EMS) involving the HEV information (computers, software, algorithms, etc.) and HEV power (power transistors, cooling system, etc.). The control system is extremely complicated and represents a tremendous challenge. By means of engaging or disengaging different clutch arrangements, the EMS can alter the structure to provide more flexibility for transmission control. The challenge is how to reduce the system complexity that involves both an electric motor and an engine for propulsion and how to coordinate these two propulsion devices to achieve optimal efficiency operation. The nontrivial decision must be made concerning the split between the motor and the engine. Once the hybrid engine torque and power are specified, the gear ratios in the transmission, the three-way gearbox, and the throttle settings must be selected to put the engine on the minimum fuel consumption line. All parts must be engaged and disengaged smoothly without jerks, shudders, shakes, or clanks. Further, perceived mismatches between driver commands and HEV responses are forbidden. Different gear ratios allow matching vehicle and wheel speed with the desired engine speed. The number of speeds in an automatic transmission affects fuel economy favorably. Multiple speed automatic transmissions begin to rival the manual shift transmission in efficiency. Speed automatic transmissions with five speeds, six speeds, and even eight speeds are beginning to appear in new cars. The ICEV engine operates from idle rpm to the maximum design rpm that yields maximum power. On the other hand, for HEVs, engines are designed specifically, having a narrower rpm operating band. Hence, the number of gears in the transmission affects the width of the engine rpm band within which the engine can be operated. The six-speed transmission gives a narrow range of engine rpm. The continuously variable transmission (CVT) has an infinite number of gear ratios between two rpm limits. CVTs are popular for hybrids because of their ability to match input/output rpms. They offer seamless acceleration and fuel economy. Locating engine operation within the rpm band for best fuel economy is only part of the solution for the hybrid control. The motor assist almost always enables operation of engine for best fuel economy by ensuring balance torques in the transmission. The key technology of the full hybrid is the electric variable transmission (EVT) system, also called the electronic-continuously variable transmission (e-CVT) system. Since the introduction of the first EVT system in 1997 by Toyota Prius, there have been many derivatives developed by different automakers. The key is to employ a planetary gear (PG) for power splitting of the engine output power, one via the ring gear to the driveline shaft while one via the sun gear to the generator, then back-to-back converters, motor, and finally the driveline shaft. On the one hand, the most efficient path for engine torque is mechanically and directly to the drive wheels. On the other hand, because of the need to balance torques in the transmission, the path of a portion of the energy is engine-generator-motor- drive wheels, which is less efficient due to the numerous transfers of energy. However, through clever use of a PG set, several components are eliminated. The PG set replaces both the three-way gearbox and the CVT. Since there are no clutches, the system involves only one physical structure that can avoid mechanical disturbances due to mode changes. Hence, under varying road loads, the engine can always operate at its most energy-efficient or optimal operation line, resulting in a considerable reduction in fuel consumption. Yet, this PG-EVT system suffers from the reliance on planetary gearing, which involves transmission loss, gear noise, and regular lubrication. In addition, the overall system is relatively heavy and bulky. Thus, there are continual research and development to solve these shortcomings, such as replacing the mechanical planetary gear by a double-rotor machine or magnetic planetary gear. The double-rotor machine is a newly introduced gearless power-split device. However, it requires brushes and slip-rings, which are less reliable and incur need for a regular maintenance. In order to overcome these drawbacks, a new class of magnetic-geared (MG) electric variable transmission systems has been developed based on contactless magnetic gears, which can offer the definite advantage of brushless and pseudo-gearless power-split operation. By replacing the planetary gearing with magnetic gearing, the resulting MG-EVT system can inherit the distinct advantages of magnetic gearing, namely the high transmission efficiency, silent operation, and no maintenance, while avoiding the use of slip rings and carbon brushes [6]. Mechanical systems are increasingly integrated with actuators, sensors, and electronics. Besides the basic energy flow in the mechanical system, an information flow in the electronic system enables a variety of automatic functions. This leads to *mechatronic systems* which consist of *mechanics* (mechanical engineering) and coupled processes (e.g., thermal, electrical), *electronics* (microelectronics, power electronics, measurement and actuator technology), and *information technology* (systems theory, automation, communication, software design, artificial intelligence). The design of the functions within mechatronic systems is performed as well on the mechanical and on the digital electronic side. Hereby, interrelations during the design play an important role and create synergetic effects. Whereas the design and the local arrangement for a conventional system are done separately, the mechanical and electronic components of a mechatronic system must be considered as an integrated overall system from the beginning. This means that *simultaneous engineering* has to take place. Adding just sensors, simple analog or digital control systems, and actuators to existing mechanical processes is not sufficient with regard to overall dynamics, robustness, space, cables, and speed of processing. Integrated mechatronic systems allow the avoidance of these disadvantages and to approach more *autonomous systems,* for example, in the form of capsuled systems. Beginning with a classical mechanical–electrical system, which results from adding available sensors and actuators to the mechanical components, one can mainly distinguish two kinds of integration for mechatronic systems [9–12]: *integration of components* (hardware integration) and *integration by information processing* (software integration). The information processing within mechatronic systems may range between simple control functions and intelligent. An intelligent control system is organized as an online expert system and comprises multi-control functions (executive functions): knowledge base, inference mechanism, and communication interfaces. The online control functions are usually organized in multi-levels: L1-control, feedback for stabilization, damping, L2-supervision with alarming and automatic protection, fault diagnosis, redundancy actions or reconfiguration, L3-optimization, coordination of subsystems, and general process management. HEV is a truly mechatronic system. The major problems with the design and control of HEV are the packaging and integration of components, and control and coordination. The design engineers have different ways to increase conversion efficiency and design a hybrid: incorporate the latest proven technology, optimize existing power train components, use new power train components, and combine and integrate optimized power train components into the hybrid. The complex power distribution system is separated into two parts at vastly different power levels and voltages. One part is the information section, which mainly comprises electronic control units (ECUs)—computers and microprocessors at low power, low voltage—which communicate by means of the control area network. Another part is the power electronics that manages the power to/from the motors/generators and to/from the battery (high power, high voltage). For a HEV, integration and control software are as important as hardware, if not more so. Combining electric motors and gasoline engines opens up many more degrees of freedom that must be controlled. Major functions of the control system are to maximize fuel economy and to minimize exhaust emissions. Minor functions of the control system include component monitoring and protection. Examples are battery state of charge (SoC), battery temperature, electrical motor overheating, and gas engine overheating. The battery merits special attention to avoid failure and to assure long life. The control system generally provides a fail-safe mode in the event of failure. Another control function, which is minor in cost but valuable in practice, is onboard diagnostics. Controlling an HEV includes essentially two sets of tasks. One is the *low-level* or *component-level* control task, where each power train component is controlled by using classical feedback control methods. The second task, referred to as *high-level* or *supervisory control*, is responsible for the optimization of the energy flow onboard of the vehicle while maintaining the battery SoC within a certain range of operation. This layer of control, called EMS, receives and processes information from the vehicle and the driver to output the optimal set points sent to the actuators and executed by the low-level control layer. The EMS also selects the best modes of operations of the hybrid power train including start-stop, power split, and electric launch [13]. The potential for optimized control is illustrated by the Honda FCX fuel cell vehicle, in which range was increased by 30% over prior versions. Of the 30% increase, 9% is attributable to an increased fuel tank size. The rest is because of a better control system [10]. This potential can be realized only with a sophisticated control system that optimizes energy flow within the vehicle. It has been recognized that the adoption of systematic model-based optimization methods using meaningful objective functions to improve the energy management controllers is the pathway to go in order to achieve near-optimal results in designing the vehicle EMS. HEVs are complex and involve many different technologies: materials, electromechanics, electrochemistry, electronics, software, and testing facilities. Among them, the motor drive technology is most actively developed in recent years where there are many innovations and advancements in the design, analysis, and control of motor drives. Motor drives are the core technology for HEVs that convert the onboard electrical energy to the desired mechanical motion. Meanwhile, electric machines are the key element of motor drive technology. The requirements of electric machines for HEVs are much more demanding than that for industrial applications (high torque density and high power density, wide speed range, high efficiency over wide torque and speed ranges, wide constant-power operating capability, high torque capability for electric launch and hill climbing, high intermittent overload capability for overtaking, high reliability and robustness for vehicular environment, low acoustic noise, and reasonable cost). On the other hand, when the electric machine needs to work with the engine for various HEVs, there are some additional requirements (high-efficiency power generation over a wide speed range, good voltage regulation over a wide speed generation, capable of being integrated with the engine).

4. State of the art in axial-flux electric machines for HEV and EV propulsion

The initial selection of electrical machines for hybrid (HEV) or electric (EV) drives includes a variety of different topologies. According to outcomes of literature survey, induction machines alongside synchronous machines take the major place in HEV or EV power trains. Both of these different families of machines topologies, in sense of operational principle, may be laid out in an axial or a radial plane. Out of radial synchronous machines, surface-mounted, inset and embedded-permanent magnet topologies, and switched reluctance machines are considered as competitive for traction purposes. Among the synchronous permanent magnet, axial-flux machines are subject of ongoing research and therefore recently the need to be included in selection process due to their advantageous axial length. The search for more efficient, cost-effective, and fault-tolerant layouts drives the design of electrical machinery up to its limits. The adoption of new materials and topologies for lightweight vehicle and power trains is extremely investment intensive. This chapter proposes to initiate the development and adoption of a new class of propulsion systems incorporating advanced electrical drives. By advanced electrical drive, it means the use of axial-flux dual mechanical output machines (the electrical machine has two independent rotors) with a single stator winding capable to control the mechanical output independently (**Figure 1**). Moreover, the requirements of many applications both in the industry and in the field of renewable energy conversion are so tough that traditional layouts are abandoned in favor of new topologies or new light is shed over older ones. We keep in mind the fact that the presence of the permanent magnet in the structure of the propulsion machine is rather a limitation which can be avoided. In order to build highly reliable propulsion systems, the use of a highly permanent magnet-depending generator/motor will limit the lifetime of the electrical machine and drive it.

Due to their high torque density capabilities, favorable aspect ratio, and the possibility to implement a large number of poles [14], axial-flux PM machines (AFPMMs) are used in many up-to-date applications. In fact, AFPMMs are applicable to fans, diesel and wind generation units, elevators, ships, and vehicles [15]. The first-harmonic approximation of the MMF field commonly used for design purposes is sometimes inadequate, for example, with trapezoidal back EMFs or with fractional slot windings. In fact, these windings operate with a high level of MMF harmonics, although they produce sinusoidal EMF. The technical literature offers many machine models, each one addressing specific issues. Some models provide a precise description of the MMF content [16], some focus on skewing [17], some provide guidelines for the use of finite element method (FEM) [18], and others account for saturation, iron losses, and temperature effects [19, 20]. Most models, however, are based on the reduction of the 3D problem to a 2D problem by the use of a cylindrical cutting plane at the main flux region

[21–25]. Moreover, 2D and 3D FEM analyses fail to account for iron losses, especially in fractional winding machines with soft magnetic composite (SMC) core [25, 26]. Some models rely on FEM simulations [27]. The influence of rotor eddy currents on the efficiency of fractional slot AFPMM is significant, as emphasized in [23]. As a result of the fundamental intricacy of the magnetic circuit, extensive papers have been devoted to the design and modeling of AFPMM. Actually, the usual first-harmonic approximation of the MMF wave is not suitable for designing windings with a fractional slot number per pole. These windings present short links, compact poles, and sinusoidal EMF, even if the MMF wave has higher- and lower-order spatial harmonics. The harmonic content does not alter the quasi-sinusoidal terminal EMF, but it alters core loss and machine inductance. The professional data available show specific models addressing certain issues: provides a detailed MMF portrayal even if the analytic model cannot be held responsible for machine curvature and presumes that the coil pitch is detailed

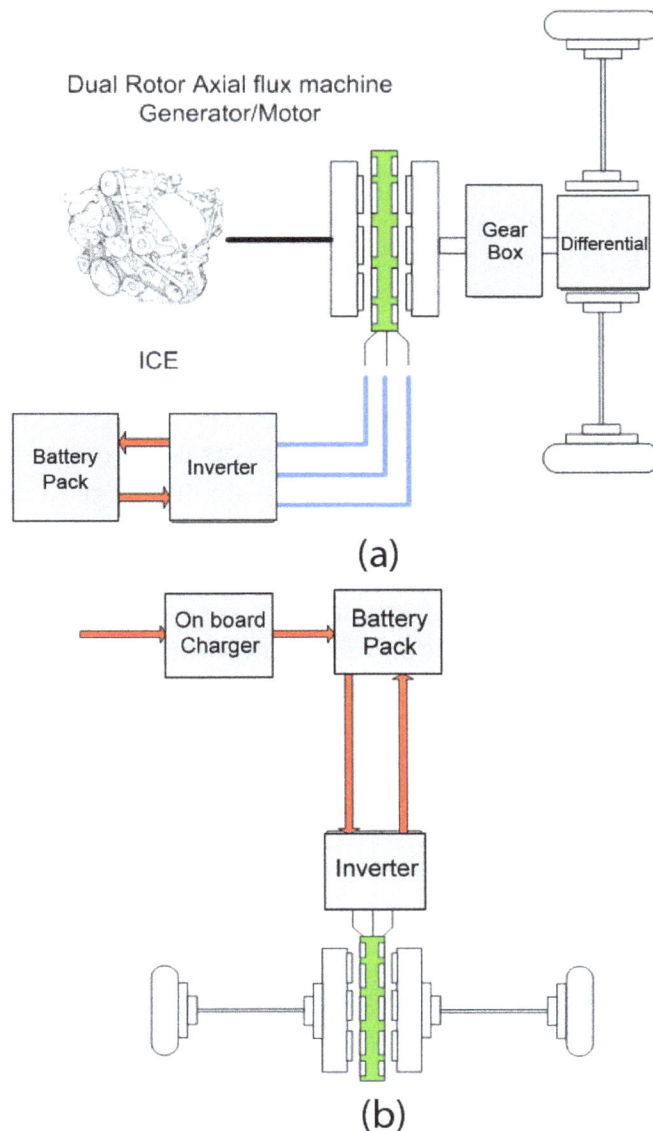

Figure 1. Lightweight power train solution based on axial-flux dual rotor. (a) Hybrid vehicle and (b) electric vehicle.

at the mean core radius, skewing-focused, aids in using the finite element method (FEM), and others are saturation, iron losses, and thermal effects accountant. However, a fair amount of models cut down the 3D issue to a 2D portrayal of a cylindrical cross section. Thermal models require a detailed analysis of rotor iron losses. Mention is based upon FEM simulations and emphasizes the impact of rotor eddy currents. The thermal transfer in axial-flux machine geometries was explained. Thermal simulations done on a machine with a massive iron rotor core showed that the rotor temperatures are above those measured over the windings surface. Actually, a considerable contribution to losses comes from the rotor and is basically due to the sub-harmonic and high-frequency parts of the stator MMF, including slotting. Recent investigations on the subject demonstrated that the sub-harmonics are, however, the main causes of rotor losses. Axial-flux induction machines are also possible. The Australian company Evans Electric has developed induction axial-flux machines for electric car as in wheel motors. These motors deliver 625 Nm and a peak power of 75 kW in an AFIM (axial-flux induction machines) configuration. This machine has the particularity that the stators cover only a part of the machine and not the complete disc. This enables a better air cooling of the machine but reduces the performance. In-wheel direct-drive motors represent the simplest and lightest method for propelling wheeled vehicles, but due to the reduced suspension performance of vehicles with increased wheel mass, the mass of in-wheel motors is a major concern. The axial-flux switched-reluctance motor (AFSRM) topology for in-wheel drive vehicle applications is presented. For a high-speed automotive generator, the axial-flux reversal machine was proposed. The high-speed AF FRM electrical machines offer great advantages of reduction in size per unit output and improved efficiency [25–27]. Small-scale generating sets of high power densities, previously used predominately in military and aircraft applications, are attracting growing attention for a wide variety of automotive applications: onboard charger, compact range extender, and turbine-based HEV. One salient feature of these sets is the use of a high-speed generator directly coupled to a small gas turbine, resulting in a significant reduction of weight and size. The path of the magnetic flux distribution in the air gap is in the axial direction. The rotor of the machine has no permanent magnet and field winding. The permanent magnets and armature windings are on the stator cores being very easy to be cooled. For this reason in the future, we start our investigation including the induction machine, the SRM machine, and the flux reversal machine. The building experience will be valuable for industry as well to the scientific community.

5. Lightweight traction motor/generator using dual-rotor single stator split shaft axial-flux PMSM

5.1. Topology description

In an effort to simplify the design of the two configurations (series and parallel), publications in the latest years released by the authors' team members, among which an international patent is present, treat a topic belonging to a complex scientific field of significant interest and certain actuality, being oriented toward identifying appropriate solutions regarding some proficient electromechanical converters and control structures of the drive systems based on axial machines excited by permanent magnets with a high number of poles, with applications in electric traction, operating in different modes: engine, generator, the brake, or combinations in between [28–35]. It proposes **an international original solution** [32] in which the two electrical machines (generator and motor) and static converters related are replaced by a single synchronous permanent magnet machine having an axial air gap, a central stator with slots on both sides, and two different windings supplied from a single PWM inverter having two output frequencies, and two independent rotors. This machine efficiency is high, and the torque density developed by the two rotors is also high. The inverter output voltage ripple contains a combination of voltages each having different frequency and determining a different rotational speed $\Omega 1$, $\Omega 2$, for the two rotors. The first rotor, coupled with the combustion engine, together with the corresponding stator winding operates as a generator most of the time (**Figure 2**) [29].

It can be used and as a starter for the combustion engine startup. The second rotor, coupled with the differential mechanism and the drive wheels together with the associated stator winding, operates as a motor (in traction) as well as a generator (regenerative braking

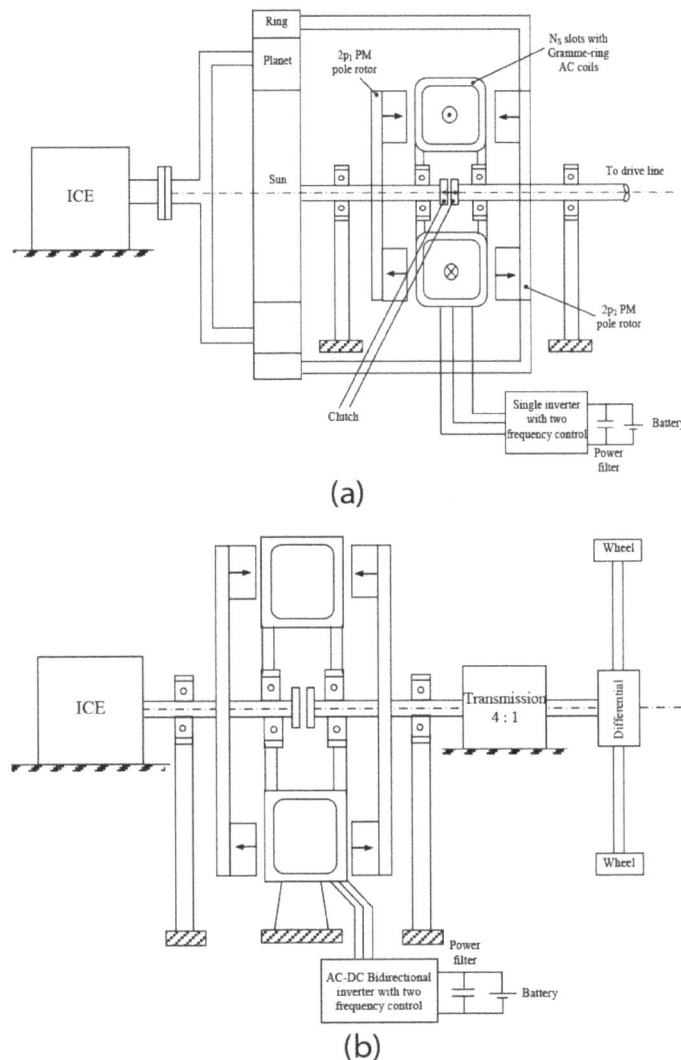

Figure 2. Principle of construction of the proposed solution: (a) parallel planetary mechanism; (b) series version [28].

regime). Fuel economy, which is obtained for the regime of urban movement, can reach 33%. For pure electric vehicles, two solutions have been found for the use of this electric drive system (**Figure 3**) where R is rotors, S is the stator, Inv. is the inverter, TM is the mechanical transmission, TD is the differential transmission, and RM are wheels.

An important advantage of using the synchronous axial air-gap single stator dual-rotor permanent magnet machine is representing the smaller length, this being able to be introduced in the clutch's place between the motor and the gearbox. A 3D drawing of the machine is shown in **Figure 4**.

5.2. Control algorithm

Traditionally, a three-phase three-leg bidirectional power converter is used in an EV. For this topology, the most appropriate power converter is the three-phase four-leg converter.

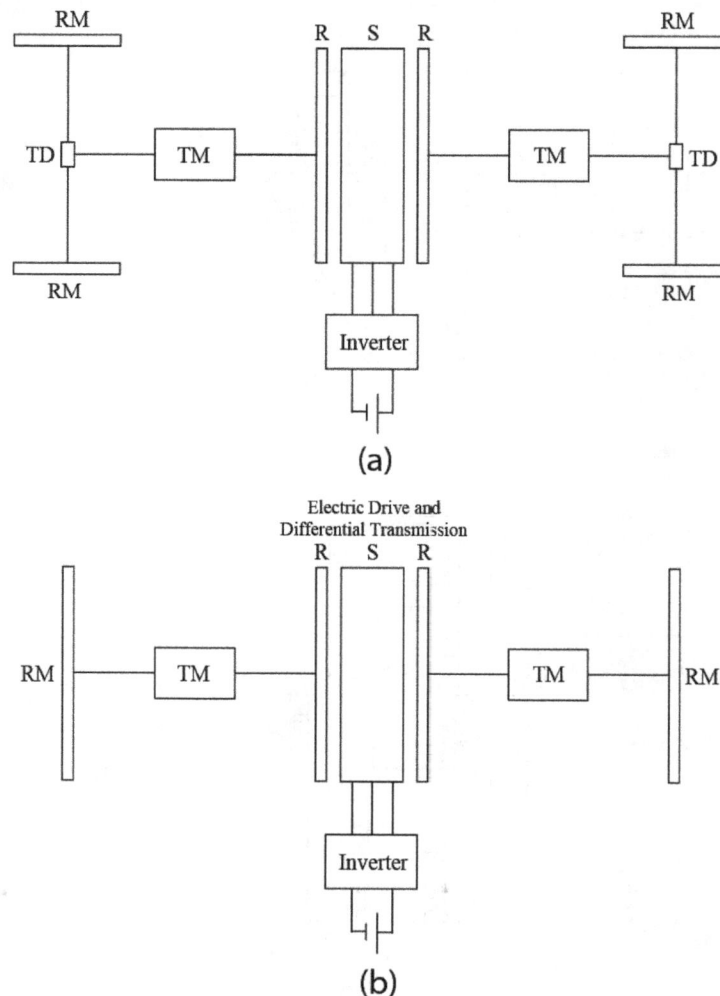

Figure 3. Using the synchronous machine with a stator and two axial rotors: (a) electric drive system for an electric vehicle with a four-wheel drive; (b) an electric drive system and a differential transmission at the same time.

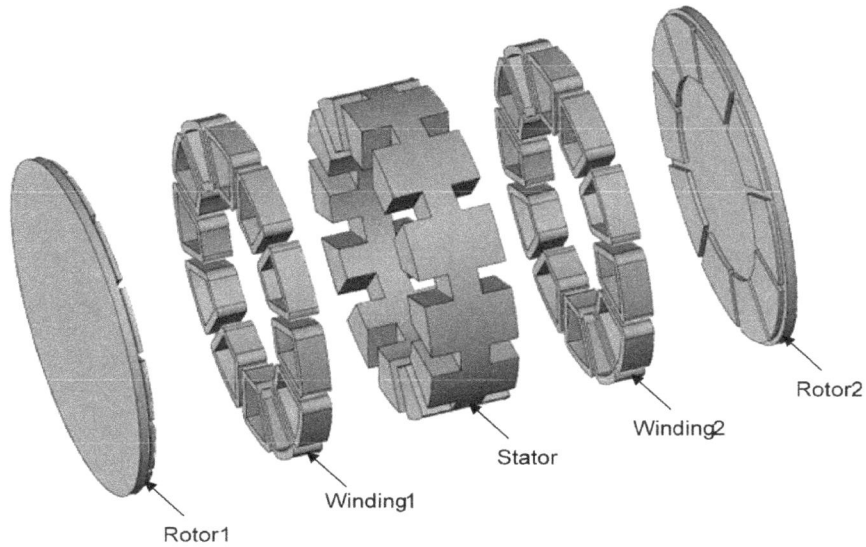

Figure 4. A three-dimensional exploded view of the proposed machine [28].

The main feature of a three-phase inverter four-leg inverter, with an additional neutral leg, is its capability to handle load unbalance. In an automotive power system, the main goal of the three-phase four-leg inverter is to maintain the desired sinusoidal output voltage over all the ranges of loading conditions and transients [34]. Typically, the EV electric machine phases have a Y-connected winding. In order to maintain the phase voltage at the same level as in three-leg inverters, one may choose a Δ connection of the two windings. This means will result in the fact that the wire diameter will be smaller, the winding may be built easier and, for an existing machine, no rewinding is needed (**Figure 5**), and in **Figure 6** for serial HEV.

Alternatively, a matrix converter solution can be used to operate the two rotating rotors (**Figure 7**). The matrix converter is an array of bidirectional switches that can directly connect any input phase to any output phase to create a variable voltage and frequency at the output.

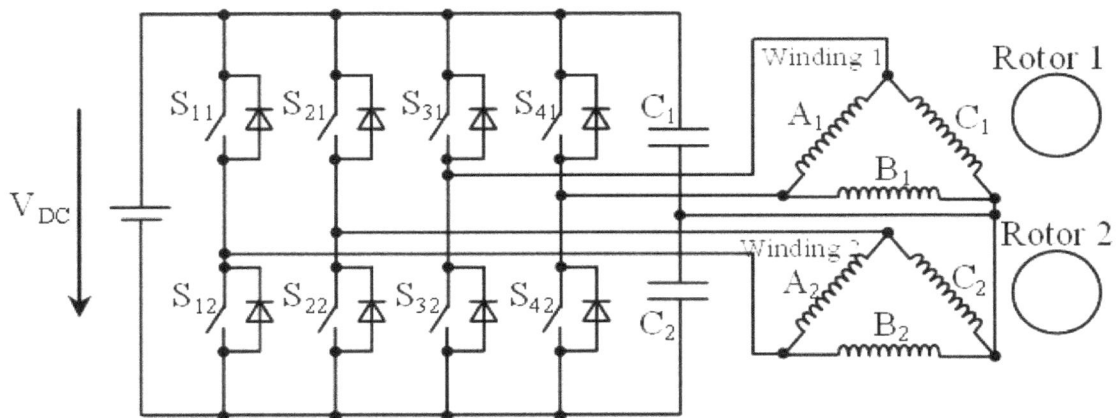

Figure 5. Three-phase four-leg inverter [28].

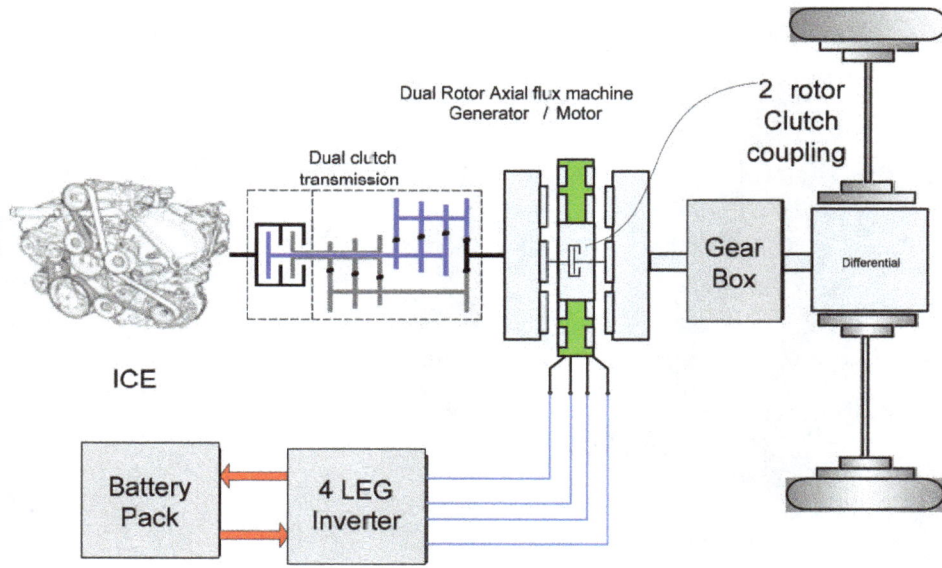

Figure 6. The four-leg inverter for serial HEV.

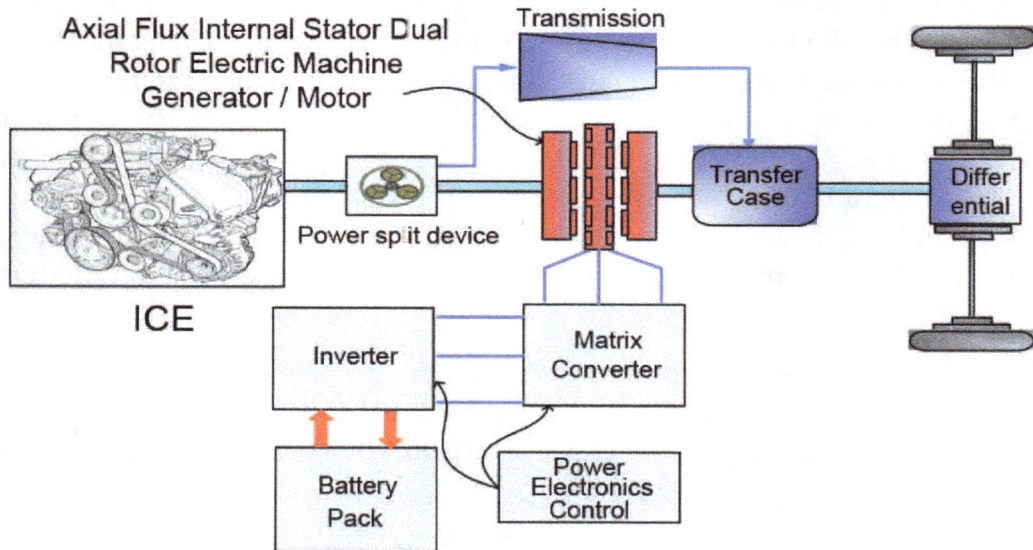

Figure 7. Parallel HEV based on matrix converter topology.

However, despite the benefit of very compact construction (no DC capacitors), this type of converter is not very easy to control, and for this reason, we only mention it as an alternative topology to be used.

5.3. Field-oriented control with four-leg inverter

The proposed configuration of a dual mechanical rotor with a single stator requires a single-phase AC current flow through the capacitors. This is due to the connection of the machine line voltages to the center point of the DC link. However, by connecting a second two-phase inverter and motor to the same DC bus, as shown in **Figure 5**, it is possible to compensate for the single-phase current in the DC link capacitors. A specially designed vector control is required for this reason. In this way, the compensation current influence on the torque is limited in the case of surface PM machines (with no saliency) [34]. The neutral voltage regulation with different current compensation is based on a simple PI controller since the motor divides the current compensation effort to both motor sides, rotor 1 and rotor 2 (in a direct ratio with their rated current) and at the same time avoids power oscillations (power pulsation occurs when two independent controllers are used). In **Figure 8**, the proposed vector control strategy is presented.

In order to evaluate the behavior of the combined solution dual-rotor single stator axial-flux PMSM machine powered by a three-phase four-leg converter, a digital simulation was performed. Using the equivalent model of the axial-flux permanent magnet machine, a Matlab Simulink model was implemented. The actual test conditions were represented by a step in the speed reference consisting of a value of 150 (rad/s) which is given for first electric machine M1 at the starting moment (t = 0) and a second step in the speed reference consisting of the value 230 (rad/s) which is given for machine M2 at t = 0.8 s (**Figure 9a**). In this scenario, the machine M1 reaches the reference speed in 0.175 s at a rated torque as load. In an EV configuration after the thermal engine starts, at t = 0.5 s, the machine M1 switches to a generator mode with 20% load at the same speed (**Figure 9b**). The speed overshooting during starting process and torque perturbation is around 2 [rad/s] (under 1.5%) which represents a very good feature of the proposed vector control. The second machine M2 is started at t = 0.8 s (**Figure 9a**), with 10% of the rated torque as load (**Figure 9b**), and it reaches the reference speed in 1.3 s. A small perturbation on machine M1 speed occurs during the machine M2 starting (**Figure 9a, b**). The machine M1 is starting with a two-time rated torque, and it runs at a rated torque between 0.2 and 0.5 s and then at 25% of the rated torque as generator. Small torque perturbation could be

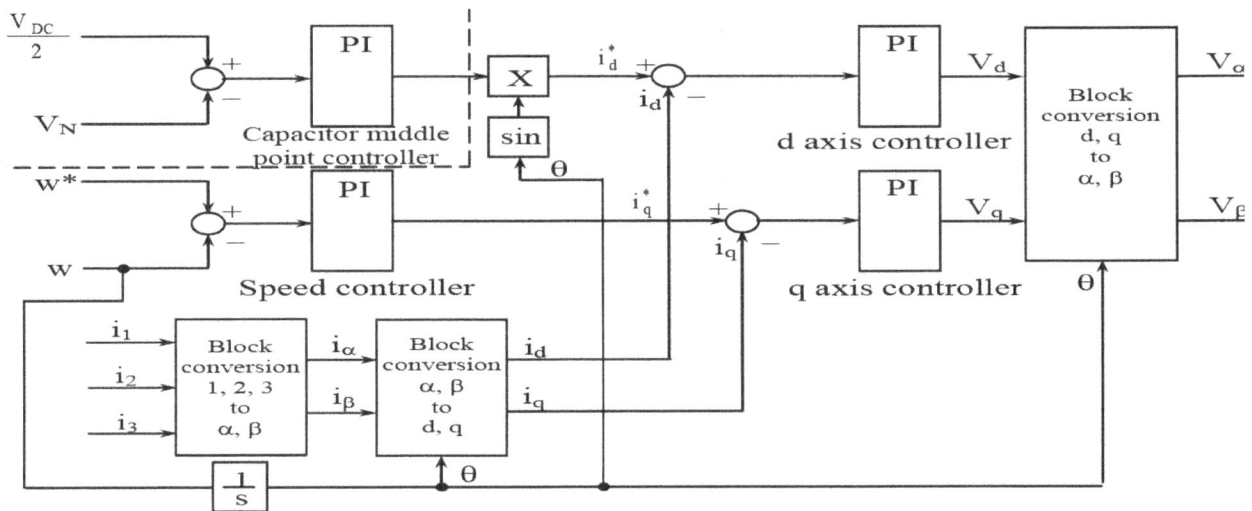

Figure 8. The proposed vector control strategy [28].

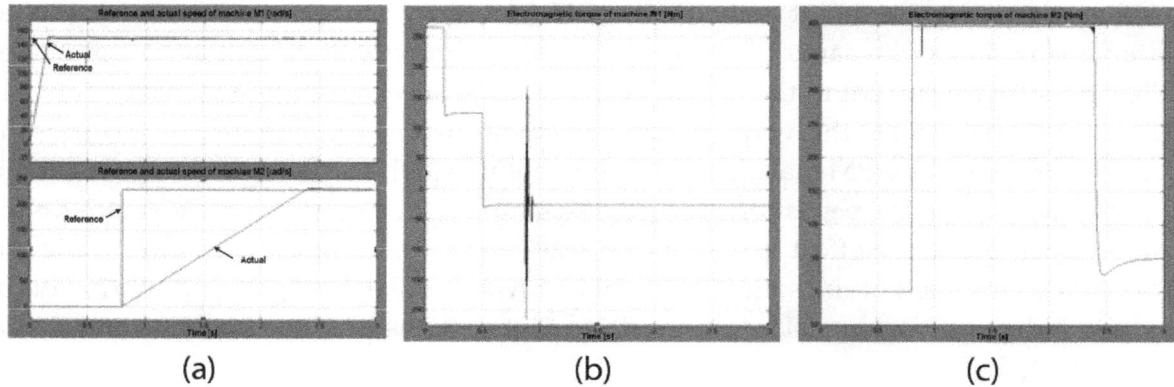

Figure 9. (a) Machine M1 and M2 reference and actual mechanical speeds; (b) achieved torque by machine M1; (c) achieved torque by machine M2 [28].

observed in machine M2 while starting. The machine M2 is starting with electromagnetically rated torque while the load torque is 10% (**Figure 9c**).

In terms of reliability, the presented topology combined with an adequate power converter allows to have the highest reliability in the single inverter configurations for EV.

6. Conclusions

The use of a very compact electrical drive will have the benefit of considerably reducing the weight and complexity of the power train. As it is known, the mechanical complexity of the power train is responsible for as much of 20% of the weight of the vehicle. The mechanical system of internal combustion engine (ICE) vehicle required a rather complex system of adaptation of speed and torque to the travel conditions. Our chapter is developed around the concept of a reduced number of mechanical elements included in the power train. This is possible by the close integration of electrical drive with the ICE and the use of the electrical differential concept. Special consideration is given to the power electronics required for the drive. Using a new approach, the number of converters is limited to one for each axle, each converter being capable to independently control the motion of the side wheels. Instead of a complex sophisticated gearbox, we propose to use a simplified gearbox or no gearbox in case of the electric differential, much of the function being fulfilled by the dual mechanical output electric machine controlled by a single power converter. A special control based on the dual vector control with operating on dual frequency will be investigated. In order to increase the ruggedness of the system, we investigate special power converters with a high degree of reliability (the four-leg converter and the matrix converter that makes no use of DC capacitors in the DC link), the multilevel inverter concept applied to EV which brings the benefit of a very reliable topology, a reduced harmonic pollution, and easy battery cell balancing. Although this seems to be an unnecessary complication to a rather proven technology, our chapter considers the fact that the existing power train solutions are not considering the problem of extra weight/complexity given by the electrification.

Author details

Sorin Ioan Deaconu[1*], Vasile Horga[2], Marcel Topor[1], Fabrizio Marignetti[3], Lucian Nicolae Tutelea[1] and Ilie Nuca[4]

*Address all correspondence to: sorin.deaconu@fih.upt.ro

1 Politehnica University Timisoara, Romania

2 Technical University, Iasi, Romania

3 University of Cassino and Southern Lazio, Italy

4 Technical University of Moldova, Chisinau, Republic of Moldova

References

[1] Ehsani M, Gao Y, Gay S, Emandi A. Modern Electric, Hybrid Electric, and Fuel Cell Vehicles. London: CRC Press; 2005

[2] Egede P. Environmental Assessment of Lightweight Electric Vehicles. Switzerland: Springer; 2017

[3] ISO (14040:2006). Environmental management—life cycle assessment—principles and framework (14040:2006) ISO Geneva

[4] Guzzella L, Sciarretta A. Vehicle Propulsion Systems. Introduction to Modeling and Optimization. Berlin: Springer; 2007

[5] Chau KT. Electric Vehicle Machines and Drives: Design, Analysis and Application. Singapore: Wiley; 2015

[6] Chan CC, Chau KT. Modern Electric Vehicle Technology. In: Oxford University Press. Oxford: Oxford University Press; 2001

[7] Njuguna J. Lightweight Composite Structures in Transport: Design, Manufacturing, Analysis and Performance. London: Elsevier; 2016

[8] Koffler C, Rohde-Brandemburger K. On the calculation of fuel savings through lightweight designe in automotive life cycle assessments. International Journal of Life Cycle Assessment. 2010;**15**:128-135

[9] Redelbach M, Klotzke M, Friedrich H. Impact of lightweight design on energy consumption and cost effectiveness of alternative powertrain concepts. European Electric Vehicle Congress, Brussels, Nov 19-22, 2012

[10] Fuhs A. Hybrid Vehicles and the Future of the Personal Transportation. CRC Press; 2009

[11] Isermann R. Mechatronics systems – with applications for cars. In: 13th Triennial World Congress, San Francisco, USA. 1996. pp. 7732-7737

[12] Isermann R. Towards intelligent mechatronic systems. Research in Engineering Design. 1996;**8**:139-150

[13] Onori S, Serrao L, Rizzoni G. Hybrid Electric Vehicles. Energy Management Strategies. Springer; 2016

[14] Cavagnino A, Lazzari M, Profumo F, Tenconi A. A comparison between the axial flux and the radial flux structures for PM synchronous motors. IEEE Transactions on Industry Applications. Nov-Dec 2002;**38**(6):1611-1618

[15] Huang S, Liuo J, Leonardi F, Lipo TA. A comparison of power density for axial flux machines based on general purpose sizing equations. IEEE Transactions on Energy Conversion. Jun 1999;**14**(2):185-192

[16] Caricchi F, Maradei F, De Donato G, Capponi F. G. Axial-flux permanent-magnet generator for induction heating gensets. IEEE Transactions on Industrial Electronics. 2010;**57**(1):128-137

[17] Caricchi F, Crescimbini F, Honorati O. Modular, axial-flux, permanent-magnet motor for ship propulsion drives. IEEE Transactions on Energy Conversion. Sep 1999;**14**(3):673-679

[18] Caricchi F, Crescimbini F, Santini E. Basic principle and design criteria of axial-flux PM machines having counterrotating rotors. IEEE Transactions on Industry Applications. Sep–Oct 1995;**31**(5):1062-1068

[19] Profumo F, Zhang Z, Tenconi A. Axial flux machines drives a new viable solution for electric cars. IEEE Transactions on Industrial Electronics. Feb 1997;**44**(1):39-45

[20] Eastham JF, Profumo F, Tenconi A, Hill-Cottingham RJ, Coles PC, Gianolio G. Novel axial flux machine for aircraft drive: Design and modelling. IEEE Transactions on Magnetics. Sep 2002;**38**(5):3003-3005

[21] Hill-Cottingham RJ, Coles PC, Eastham JF, Profumo F, Tenconi A, Gianolio G. Multi-disc axial flux stratospheric aircraft propeller drive. In: 36th. Proc. IEEE Industry Applications Society Annual Meeting. Oct 2001

[22] Parviainen A, Niemela M, Pyrhonen J. Modeling of axial flux PM machines. In: IEEE International Electric Machines and Drives Conference, 2003. IEMDC'03. Vol. 3. Jun 1-4, 2003. pp. 1955-1961

[23] Bumby JR, Martin R, Mueller MA, Spooner E, Brown NL, Chalmers BJ. Electromagnetic design of axial-flux permanent magnet machines. IEE Proceedings – Electric Power Applications. Mar 2004;**151**(2):151-160

[24] Abdel Karim N, Azzouzi J, Barakat G. Influence of skewing on the performances of an axial flux PM wind generator coupled to a diode rectifier. In: IEEE International Electric Machines & Drives Conference. 2007. IEMDC '07. Vol. 2. May 3-5, 2007. pp. 1037-1042

[25] Kano Y, Tonogi K, Matsui N. A simple non-linear magnetic analysis for axial-flux permanent magnet machines. IEEE Transactions on Industrial Electronics. June 2010;**57**(6): 2124-2133. DOI: 10.1109/TIE.2009.2034685

[26] El-Refaie AM, Jahns TM, Novotny DW. Analysis of surface permanent magnet machines with fractional-slot concentrated windings. IEEE Transactions on Energy Conversion. 2006;**21**(1):34-43

[27] Gieras JF. Axial Flux Permanent Magnet Brushless Machines. Dordrecht: Kluwer Academic Publishers; 2004. 340 p. ISBN: 1-4020-2661-7

[28] Tutelea LN, Deaconu SI. Dual Rotor Single Stator Permanent Magnet Motors for Hybrid Electrical Vehicles. Timisoara: Politehnica Publishing House; 2014

[29] Boldea I, Topor M, Marignetti F, Deaconu SI, Tutelea LN. A Novel, Single stator dual PM rotor, synchronous machine: Topology, circuit model, controlled dynamics simulation and 3d fem analysis of torque production. In: OPTIM 2010, Brasov, Romania. May 20-22, 2010. pp. 343-351

[30] Tutelea LN, Deaconu SI, Boldea I, Marignetti F, Popa GN. Design and control of a single stator dual PM rotors axial synchronous machine for hybrid electric vehicles. EPE 2011, Birmingham, England. 30 Aug–2 Sep, 2011

[31] Tutelea LN, Boldea I, Deaconu SI. Optimal design of dual rotor single stator PMSM drive for automobiles. In: International Electric Vehicle Conference, Mar 4-8, Greenville, SC, USA, 2012, p. 8

[32] Boldea I, Deaconu S, Marignetti F, TuteleaL. Brushless electrical actuator with two independent rotors for hybrid electrical propulsion. Patent Number: IT1409332-B, 2014. 2014-Q06615

[33] Boldea I, Tutelea LN, Deaconu SI, Marignetti F. Dual rotor single – stator axial air gap PMSM motor/generator drive for HEVs: A review of comprehensive modeling and performance characterization (ESARS). 2012 Electrical Systems for Aircraft, Railway and Ship Propulsion, Bologna, Italy

[34] Tutelea LN, Boldea I, Deaconu SI. The single stator dual rotor PMSM for HEV: Two windings and 4 leg inverter control. In: EPE PEMC 2012 ECCE Europe, Novi Sad, Serbia. Sep 4-6, 2012

[35] Tutelea LN, Deaconu SI, Boldea I. Design and FEM validation for an axial single stator dual rotor PMSM. IECON 2012 – 38th Annual Conference on IEEE Industrial Electronics Society, Montreal, Canada. Oct 22-25, 2012

4

Electric Vehicles Integrated with Renewable Energy Sources for Sustainable Mobility

Michela Longo, Federica Foiadelli and Wahiba Yaïci

Abstract

Across the globe, governments have been tackling the concerning problem of air-polluting emissions by committing significant resources to improving air quality. Achieving the goal of air purification will require that both the private and public sectors invest in clean energy technology. It will also need a transition from conventional houses to smart houses and from conventional vehicles to electric vehicles (EVs). It will be necessary to integrate renewable energy sources (RESs) such as solar photovoltaics, wind energy systems and diverse varieties of bioenergies. In addition, there are opportunities for decarbonisation within the transportation sector itself. Paradoxically, it appears that the same transportation sector might also present an opportunity for a speedy decarbonisation. Statistics indicate that transportation is responsible for 14% of global greenhouse gas (GHG) emissions. However, there are numerous options for viable clean technology, including the plug-in electric vehicles (PEVs). There are indeed many technologies and strategies, which reduce transportation emissions such as public transportation, vehicle light weighing, start-stop trains, improved engine technology, fuel substitution and production improvement, hydrogen, power-to-gas, and natural gas heavy fleets. This work concentrates on EV adoption integrated with RES. Specifically, this chapter examines the feasibility of significantly reducing GHG emissions by integrating EVs with RESs for sustainable mobility.

Keywords: electric vehicles (EVs), renewable energy sources (RESs), solar photovoltaic energy, wind energy, vehicle fleet, smart grid, pollutant emissions

1. Introduction

The transport sector is the main responsible of the air pollution in European cities, as it produces almost a quarter of all the greenhouse gas (GHG) emissions. There has been a decrease in the emissions since 2007, but they are still remaining higher than in 1990. Road transport, in particular, was considered responsible for more than 70% of GHG emissions from the transport sector in 2014 as it can be observed in **Figure 1** [4].

In December 2015, 195 countries including Canada acceded to the Paris Agreement, an additional international measure to more increase efforts to address climate change through a reduction of global GHG emissions, and restriction of the global average temperatures to below 2°C. According to the Intergovernmental Panel on Climate Change, this is based on understanding that 2°C is the maximum allowable emissions threshold, after which irreversible climate harm would have occurred. All efforts must consequently be on deck to resolutely instigate to lower GHG emissions in order to avoid unsustainable climate conditions from occurring [1–3]. Considering that nations have different sectors contributing to their respective GHG emission profiles, each is anticipated to accomplish its obligation by engaging diverse procedures. For example, Canada, which signed its official commitment in April 2016, must somehow reduce GHGs by 30% below 2005 levels by 2030, and this turns to reducing emissions between 200 and 300 megatonnes from projected levels.

The European Commission has adopted a low-emission mobility strategy, a global shift towards a low-carbon, and circular economy since 2016. The European strategy for the transport sector consists in an irreversible shift to low-emission mobility, as it is mandatory to reduce air pollutants critical for our health. GHG from transport will have to be at least 60% lower than in 1990 and be firmly on the path towards zero. The strategy integrates a broader set of measures to support Europe's transition to a low-carbon economy and supports jobs, growth,

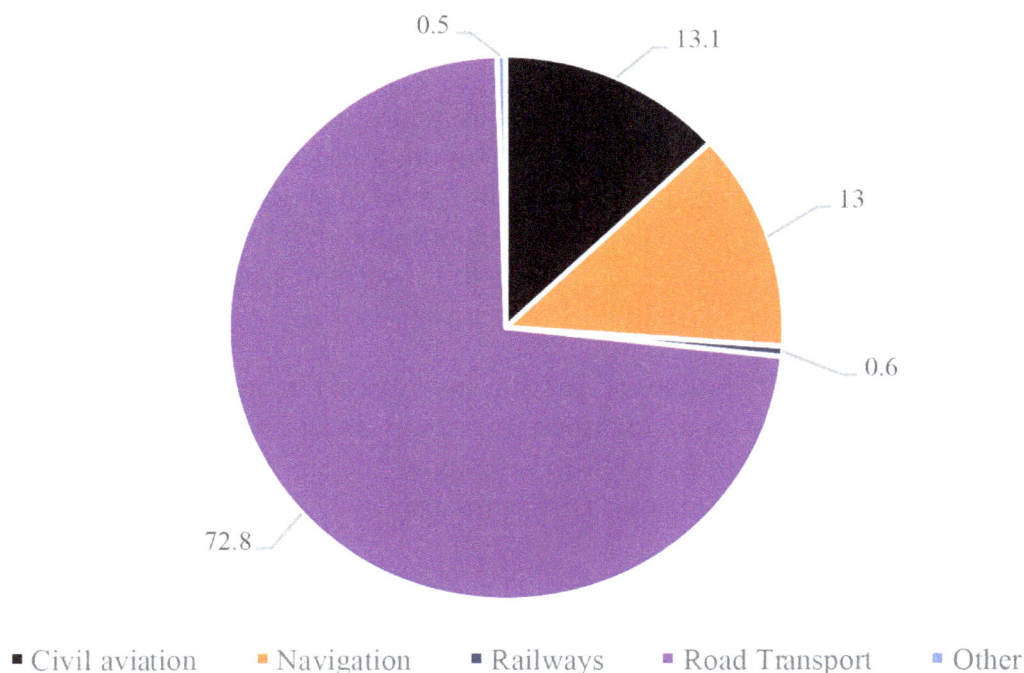

Figure 1. GHG emissions from transport by mode in 2014 [4].

investment and innovation. The strategy will benefit European citizens and consumers—by improving the quality of the air, reducing the levels of noise, lowering the levels of congestion and ameliorating the safety [4].

The measures to be taken to accomplish air purification will include:

- leveraging digital technologies, implementing affordable pricing and promoting the move to decrease emissions from transportation, in order to enhance the efficiency of the transport system;

- encouraging and accelerating a revolutionary shift to internal combustion engines, and alternative sources of energy for transportation having lower emission levels and which use alternative fuels such as hydrogen, innovative biofuels and renewable synthetic fuels, and electricity;

- accelerating the conversion in the direction of low- and zero-emission vehicles.

A critical determinant of the success of these strategies is the consistent support of local authorities. These authorities may offer incentives to people to use vehicles with low-emission based on their employing alternative energy sources. They may also encourage the use of other methods of conveyance, including biking and walking, public transport and car-sharing/pooling arrangements, which effectually decrease polluting emissions. In addition to the European Fund for Strategic Investment, EUR 70 billion is offered for transport under the European Structural and Investment Fund, comprising EUR 39 billion for funding the change to low-emission mobility, of which EUR 12 billion for low-carbon and sustainable city mobility only. A fund of EUR 6.4 billion is offered for low-carbon mobility projects in the research programme Horizon 2020.

In the last decade, EVs have become in some way widespread, principally because of their negligible flue gas emissions and lesser reliance on oil. It is estimated that by 2022, EVs will be over 35 million in the World. However, a critical problem associated with EVs is that their high penetration raises branch and transformer congestion and heavy electricity demand to the power grid. One efficient approach to relieve the effect is to integrate local power generation such as RESs into the EV charging infrastructure [5–8].

There is a lack of systematic studies considering the interaction and integration of EVs with renewable RESs, the power grid, the charging infrastructure and the strategies to decrease air pollution, all together. This chapter seeks to investigate the integration of EVs with RESs for sustainable mobility in greatly reducing GHG emissions.

The structure of this chapter is as follows: in Section 2, a description of solutions for reducing CO_2 emissions in road transportation is provided. Section 3 discusses in detail the EV interaction and integration with RESs including wind energy, solar energy, the electrical network integration with the management of distribution grid, and electric vehicle coordination; while the conclusion is presented in Section 4.

2. Road transportation and the menace of harmful emissions

Road transportation generates about one-fifth of the EU's overall emissions of CO_2, the principal GHG. Although these emissions reduced by 3.3% in 2012, they are nevertheless 20.5% greater than that generated in 1990. Transport is the main sector in the EU where GHG emissions are still increasing. Europe can achieve its long-term transition to a low-carbon economy if it transforms its road transport sector. EVs powered with electricity from RESs can diminish future air pollutant emissions, including GHGs from road transport. It is found that 15% of the EU CO_2 emissions are generated by light-duty vehicles and falling each year as the automotive industry works to achieving EU emission targets. Member states are required to disseminate relevant information to drivers. The car and van targets for 2015 and 2017, respectively, were attained in 2013.

EU legislation obliges member states to certify that appropriate figures are delivered to users in order to guide drivers' choices of vehicles with low fuel consumption and that vehicles must display labels indicating a vehicle's fuel efficiency and CO_2 emissions.

Trucks and buses are accountable for 25% of CO_2 emissions from road transport in the EU and for approximately 6% of total EU emissions. In spite of some amelioration in fuel consumption efficiency in recent years, these emissions are continually growing, essentially owing to increasing movement of road cargo. To mitigate these issues, the European Commission is currently developing an elaborate plan to decrease CO_2 emissions from heavy-duty vehicles.

It is noteworthy that fuel quality is an essential factor in decreasing GHG emissions that emanate from transportation. EU regulation imposes that the GHG concentration of vehicle fuels to be reduced by up to 10% by 2020.

Electricity instead of oil for vehicle propulsion will contribute to achieve the European Union targets on CO_2 emissions reduction. The electricity needed could be produced by various renewable and carbon-free energy sources. In fact, the EVs have a three times higher efficiency than internal combustion engine vehicles. Moreover, they emit no tailpipe CO_2 and other pollutants like nitrogen oxides (NO_x), non-methane hydrocarbons (NMHC) and particulate matter (PM). Furthermore, they are really silent and do not produce any vibration. The future optimisation of EVs is focused on technological optimization and market development. On the technology side, the main efforts are on the reliability and durability of batteries and supercapacitors, on the reduction of battery weight and volume, on improving their safety and on reducing their cost. Other technological challenges regard improving hybrid electric powertrains, charging infrastructure and plug-in solutions.

The European Commission promoted a Europe-wide electromobility initiative known as Green eMotion. In this plan, EUR 41.8 million invested by 42 partners both public and private within the energy sector and supported with 24.2 from the European Commission with the goal of exchanging and developing knowledge and experience, and enabling the deployment of EVs in the market. However, currently, there are other ongoing and challenging projects with focus on mobility, in particular eMobilita (electromobility in urban transport: a multi-dimensional innovation (socio-economic and environmental effects)) [9], NEMO (hyper-network for electromobility) [10] and OSEM-EV (optimised and systematic energy management in electric vehicles) [11].

The potential of RESs to power EVs can help reduce pollution, with a considerable decarbonisation effect and improve resource efficiency. Surely, it varies a lot by country, based on the level of the infrastructures and on the demand for additional electricity. It is likely that additional electricity generation will be needed in the European Union to cater for the extra demand arising from approximately 80% share of EVs in 2050. According to the recent report, it is estimated that the European electricity consumption from electric cars will shoot from roughly 0.03 in 2014 to 4.5% by 2030 and approximately 9.5% by 2050. It is highly argued that additional power production is the ultimate way of meeting the other electricity demand resulting from the high rates of EVs ownership. Moreover, the extra electrical energy needs to be incorporated in the infrastructural system of Europe. Currently, it is quite impossible to tell, how much electricity is required and what type of production can be sufficient to cover the current electricity demand. However, once certainty has been attained, the increasing demand for electricity generation is likely to have an enormous impact on the overall power system in Europe. It is estimated that the electricity consumption needed by an 80% share of EVs in 2050 will differ between 3 and 25% of total electricity demand across the member state of EU. According to the department of dynamic management, an additional electrical capacity of about 150 GW will be required to charge the traditional electric car. In general, the increased number of EVs will automatically need an additional generation of electricity. As for the nations with a similar share of renewable energy, the management strategies may differ in the attempt of accommodating the charging of the increased number of electric cars. Importantly, the core principles of management strategies depend on the nation's types of renewable energy as well as conventional power generation systems. For instance, states characterised by high solar energy production capacity for which the preferred charging peak will be during the day will have to adopt different power management strategies from countries which only depend on wind, combined solar or wind electricity production. In addition, it will be necessary for regions with weak network infrastructure to add grid reinforcement or rather implement specific smart charging approaches to ensure efficiency as well as flexible electricity production and distribution infrastructure. The main benefit related to increasing the number of EVs is that it significantly minimises direct emissions of CO_2 as well as air pollutants from road transport. Nonetheless, these positive impacts could be partially offset due to additional emissions caused by increased amount of electricity needed as well as continued use of fossil fuel in the power industry. That is to say, lower emissions of CO_2 and air pollutants determined by a substantial increase of EVs could cause higher emissions by the electricity generation when it is based on fossil fuel combustion and when the reduction in electricity demand is not made in other sectors.

Overall, the avoided CO_2 emissions in the road transport sector should outweigh the higher emissions from electricity generation. In countries with high shares of fossil fuel power plants, electric vehicle demand could, however, lead to higher CO_2 emissions. Ecologically, the significance of electrically driven automobiles in cases of EV aggregator may not wholly be accomplished. As a case example, 80% portion of electrically operated vehicles in about 2050 can help reduce the unswerving exhaust emissions released by each contaminant by roughly a figure higher than 80% as compared to the levels in 2010. Nevertheless, concerning carbon dioxide gas, the total reduction of nitrogen oxides, as well as PM, may to some extent offset the further releases of the toxic pollutants resulting in the electric power production sectors. The reduction depends on the type of contaminant, for instance, 1% for NO_x while 3% for PM_{10} (PM_{10} refers to the particulate nature of matter). Unfortunately, the situation is different for sulphur compounds. Furthermore, sulphur oxides are released into the environment in various ways, for example, through the exhaust fumes from the fuel driven automobiles and the use of coal as a source of energy. These emission entries will additionally result in further pollution whose results will exceed by 5% the mitigation capacity designed by the road transport department. Notably, the extra abatement strategies of the highly concentrated SO_2 emissions are a need that cannot be negotiated.

The main disparity regarding air contaminants resulting from road transport segment and the power production is directly incomparable based on their corresponding effects on humanity. Systematically, these impacts rely on a more prominent extent on locality, the intensity as well as types of emission sources. Pollution from automobiles happens at the ground level and usually, in such areas as residential, workplaces, offices among others. In comparison, contamination is significant within the cities, municipalities as well as towns. A considerable portion of the population is exposed to pollution. Contrastingly, power plants are built outside the cities/towns where there is sparse population. Since there is low-exposure, there is a change in the release of contaminants, for instance, from roads to power generation sector, which is a very significant environmental concern for public health.

A considerable portion of chargeable cars primarily on the European highways in the future is anticipated to have consequences on electricity production as well as distribution infrastructures. In the process of incorporating extra electricity need, it is worth considering that road and energy sectors should be firmly joined. Besides, the decisions regarding policies as well as investments across these two sectors should be integrated. Chargeable automobiles are just another way through which not only the European continent but also the entire world can make positive steps forwards towards the achievement of a more sustainable as well as resource-efficient economy as well as carbon-free transportation structure. The replacement of the conventional cars with electricity-driven ones is a significant means of reducing emissions, but this relies on which source of electricity to rejuvenate the vehicles. Causes may include; renewable sources, fossil fuels and also nuclear. Therefore, blindly substituting the conventional cars is not a perfect solution to the transport-related issues, for instance, rapid congestions as well as demand for road transportation. It is evident that proper systems of the transport are urgently required, and this may encompass additional expansion of renewable energy like biofuels. In a nutshell, a deviation in the public means of transportation as well as the underlying structures. Undoubtedly, this approach will help accomplish the EU obligation to ensure an eco-friendly economy [12].

3. Electric vehicle integration with renewable energy sources (RESs)

The growing wind power and solar photovoltaic (PV) installed capacity has initiated high-requirements on power balance control and power quality in several regions in Europe [13–15].

Large offshore wind farms tend to direct a high-power capacity at a single location. Notably, the magnitude of the power fluctuation can reach extremely high values due to wind speed variations. For instance, the 10-min average wind power profile from the 160 MW wind farm Horus Rev. is depicted. Within 11 days, the normalised wind power generation differs from zero to approximately 100% production. Wind power fluctuations are visible at different time scale such as short (intra hour) and long (several hours) [6, 16–28]. Furthermore, wind power and PV are known as non-dispatchable energy sources since the active power production is variable over time [6, 17–29]. The transmission system operator (TSO) plays a significant role in ensuring the balance between consumption and production at all times including at intra-hour time scale.

Importantly, the research has majorly focused on the potential contribution of EVs to facilitate the integration of RESs in the power system. The research topic has been developed within the paradigm of the smart grid, and it is centred on the EVs potential of establishing mutual benefits to both the electric power system with RES and future EV users. The term "smart grid" refers to the operation of the power system using communication, control technology, power electronics technologies as well as storage technologies to balance production and consumption at all levels [18]. Regarding this vision, an EV is in a position to act as a controllable load or as storage, charging or discharging part of its battery capacity back to the grid, conferring to the vehicle to grid (V2G) notion, as stated by Kempton and Letendre [19].

If the charging of EVs is uncoordinated, their impact on the grid is equivalent to a large electric load resulting in higher power systems peak-load and to distribution grid congestion issues [20]. To avoid such scenarios, the study has researched on what impacts EV coordinated charging can have in correlation with RES production. To be precise, the research has focused on the solutions using EVs for the provision of ancillary services for wind integration as well as energy storage for PV integration.

3.1. Electric vehicle integration with wind energy

Using EVs in power systems together with wind power has been reported to be ideally suited for the provision of ancillary services [18]. According to Kempton et al. [21, 22], EVs should be coordinated for high-value services including ancillary services that often reduces the operating cost to EV owner in the short-term period. The EV owners are likely to experience lower price despite a higher initial value compared to the ICE cars. Pillai and Bak-Jensen [23] examined the benefits of ancillary services provisions by EVs in the western Danish power system. They mainly checked at the disposition of secondary reserves, load frequency; control (LFC), which is assessed through simulation models. The authors explain how EVs can efficiently control power mismatch resulting from the variability of wind power, therefore eliminating the use of conventional power plants. Galus et al. [24] developed a similar idea where large amounts of clustered EVs, as well as household apparatus, can be used to offer secondary reserves, LFC in the power system. Regarding the simulation results, it is observable that vehicles batteries are subject to extensive energy excursion, which tends to pass from empty to full state of charge. The report by Hay et al. [25] analyses the opportunity of using EVs for regulating power: it is believed that in order to reduce the excessive use of automatic reserves [26] and in order to re-establish the availability of these, it is necessarily essential to increase the purpose of regulating power in Denmark. The authors affirm that using EVs to offer monitoring power in Denmark is one of the most effective solutions for substituting the reduced reserve power generated from conventional power plants in the future. The research also looked at micro-grid applications. In [27, 28], Lopes et al. discuss how EV load coordination can be used in stabilising the system frequency in micro-grid with wind power by using a droop control strategy. It was discovered that the penetration level of wind power could be increased even more by using a coordinated EV load. In a different study, EVs and power systems are believed to be perfect as controllable loads in simulation environment; however, the research failed to address the possible hindrances like EV control requirements as well as EV elements response during moments of coordination.

3.2. Electric vehicle integration with solar energy

The research on the usage of solar power through EVs is significantly diversified as compared to various studies focusing on wind power and EVs. Substantially, it is possible to generate electricity PV at both medium and low voltage levels within the power systems. Besides, this alternative additionally motivates the concept of incorporating the PV generation with EVs [6]. As evident from the analysis conducted by Bessa and Matos [29], utilising EVs in the process of distributing grids using PV is considered an alternative to storing energy instead of controllable loads vis-à-vis the suggestions made by various scholars. Additionally, research reveals that during the daytime when the solar radiation is at the peak, solar power can be easily stored in the car batteries for future usage. In this field, several contributions, for instance, the idea of "green" charge [29] has enabled people to understand the significance of maximising the cost of EVs throughout the irradiation period. On the contrary, Birnie [30] also recognised another application. The scholar introduced a concept in which EVs can be charged during the day at the parking areas situated, for instance, within the workplaces. In addition, EVs can be re-energised entirely during the working periods to realise the solar-to-vehicle (SV2) approach. Furthermore, the research also illustrates that energy generated in each parking area is essential in the extra generation of adequate electricity for transportation requirements for the EVs operator.

Although grids may have high penetration capacity of PV, they may too have lower voltages. In such situations, the primary constraint is linked to variations of voltage magnitudes along the feeders. Moreover, such discrepancies can be noticed particularly within the periods of high production as well as the conditions of low load [31–34]. Unsurprisingly, these events are likely to occur regularly, but on areas majorly the places of residence that have highly concentrated roof-top PVs. Conversely, many studies have examined various alternatives in the mitigation of voltage capacity, for instance, the grid reinforcement [32], approaches to reactive power control [34–38], harmonised active power curtailment [33], as well as permanent storage of energy [39].

Research exposes that the application coordinated EV load within the feeders using high PV penetration has not been satisfactorily analysed.

3.3. The electrical network integration with the management of distribution grid

The need for a transition to a more sustainable energy system leads to a deep change in the energy, building and transports sector. Power installation from RESs is becoming more and more relevant, new mobility schemes, namely car sharing, are growing more popular and particular attention is paid to energy efficiency in buildings. Moreover, each of these aspects is related to another important concept that is energy storage. The greatest change in the energy sector has occurred due to the development of distributed (or diffused) generation (DG). According to DG consists of the totality of power plants having a nominal power lower than 10 MW and connected to the distribution network. DG plants exploit primary energy sources—in the majority of cases renewable—which are distributed on the territory (thus the name distributed generation) and that could not otherwise be exploited in a traditional centralised plant; they supply local loads and they can be operated in a co-generative mode. In an urban district, examples of DG are PV panels and solar collectors mounted on top of buildings.

One of the drawbacks of DG is the high specific investment cost mainly due to the fact that, being medium or small plants, scale economy cannot be applied. Nevertheless, this can be faced thanks to a suitable incentive strategy. The real problem is the difficulty in predicting and controlling the power produced and put on the distribution network. So the distributed generation, together with other distributed energy resources such as EVs and energy storage, is the main driver for the shift to a new paradigm in the management of the grid: the passage to a smart grid.

Smart grid is defined as a modern electric power grid infrastructure that guarantees the reliability of the system and the security of supply, allowing to face problems related to the distributed power generation from RESs and to control the load, promoting energy efficiency and involving the passive final users. In order to do so, integration of the electrical grid with information and communication technology (ICT) is needed.

The availability of electricity is of crucial importance for all human activities. Therefore, continuity and security of the supply service are necessary. Since nowadays electrical energy cannot be stored at low-cost and in large quantities, electrical systems must guarantee a constant equilibrium between production and consumption. This means that the power generated has to correspond exactly to the one requested at any time interval. The electrical network is ruled in a way to ensure that this balance is respected despite any possible disturbance, from load fluctuations to faults determining the unavailability of some grid elements. The structure of the electrical grid is investigated in the paragraph below.

In a traditional network configuration, power generation occurs mainly in big, centralised power plants. The energy produced is put on the transmission network at high voltage (HV network, 132–220–400 kV), then it is transformed into medium voltage (MV network, 15–20–23 kV) through primary substations and eventually it is converted in secondary substations into Low Voltage (LV network, 230–400 V) and distributed to the final users. Domestic and commercial users are generally connected to the LV network while the majority of industrial users are connected to the MV network.

HV grid is designed to transfer bulk power from major generators to areas of demand; it has a network structure to ensure different alternative paths for the power flow in case some of its elements are unavailable owing to a fault. In Italy, for example, the transmission network has reached a high automation level, leading to a good reliability and security of supply. The transmission grid is ruled by the transmission system operator (TSO) which in Italy is Terna.

On the other hand, MV and LV grids have a radial structure. To be more precise, even if in MV grids different possible paths are available for the power flow, these grids are operated in a radial way. Distribution networks (MV and LV grids) are managed by the distribution system operator (DSO). Distribution networks have originally been conceived to transfer power just in one direction: from the substations to the final customer. This model is appropriate as long as only loads, with the exception of a very few generators, are connected to distribution networks so that they can be regarded as passive. Due to recent large spreading of distributed generation, mainly from renewables, this model needs to be re-visited.

At present, generators are connected to the distribution network according to the fit and forget approach. In the fit phase, the DSO checks that technical rules for connection to the grid are respected and that the generator's functioning does not create any problem in any credible operational scenario. The capability of the distribution grid to accept a certain amount of diffused generation is called hosting capacity. The nodal hosting capacity (NHC) methodology is utilised to determine how much DG can be connected to a given network respecting performance limits. Typically, the operating limits are as follows: rapid voltage changes, short circuit currents, reverse power flow and line thermal limits. The DSO deals with the generator as a negative load, since; it puts power on the network instead of withdrawing it. So, the DSO is forgetting it because it cannot control the generator during its operation: The generator can introduce power into the grid at any moment depending on the will of the producer or on the availability of the energy resources. Therefore, it is possible to identify three main issues related to the actual distribution network. First of all, the DSO is obliged to limit DG connection in order to keep the control of the grid, reducing the power that could be installed through DG. A way to overcome this issue would be allowing the generators to collaborate to the management of the grid. The second drawback is related to the behaviour of DG in case of faults or contingencies. If there is an anomaly in the measured values of frequency and voltage at the connection point, DG is disconnected even if the problem is not related to the distribution network but to the transmission one. This results in the sudden unavailability of the power produced by DG which could have dangerous consequences for the safety of the overall electrical power system.

Last but not least, reverse power flow can occur if the installed DG grows. It means that the power does not flow anymore only from the substations to the users but vice-versa. Thus, the updating of protection and regulation systems is necessary. Provided all this, a transition to a new management of the electrical network and of the entire power system is needed; we refer to this new model as smart grid. The aim is to move from a system in which power production is centralised and controllable while consumption is completely random, so the responsibility for the balance between generation and consumption is entirely on the production side, to a system in which part of the generation is non-programmable, but this can be counterbalanced by a controllable portion of consumption. The idea is to move from a passive grid to an active one in which there is a bi-directional exchange between producers and users as it shown in **Figure 2**.

So, besides all the technical aspects regarding protections and regulations, in order to face the changes brought in by DG, the involvement of the final user has a fundamental role because he/she has to become responsible for part of the load control.

While EVs expend power, they can likewise send out energy to the grid as 'mobile energy' units of storage. An expansion in electric vehicle reception may mean greater adaptability for the framework to react to supply as well as demand. The vitality providers will offer the "vehicle-to-grid" administration to purchasers of the EVs. For this situation, reserve funds from vehicle-to-grid administrations could take care of the yearly expense of charging an electric auto. The proprietor of the auto should introduce a unique charger at home and the provider will deal with the auto's battery. The energy providers will naturally exchange power from the auto's battery to the grid amid peak times when costs are most noteworthy, possibly giving money related come back to the proprietor. The auto's battery will be charged amid off-peak times when costs are least. This improvement shows that vehicle-to-grid administrations could be utilised as a part of different nations amid peak hours to discharge stored

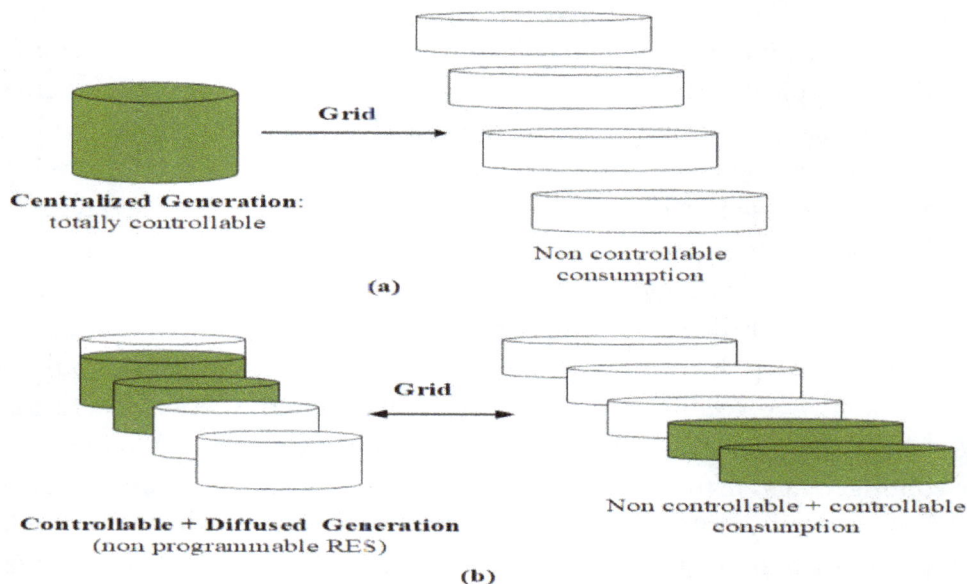

Figure 2. Comparison between the (a) actual and the (b) future electric system.

power onto the grid and reduce charging costs. Adjusting the supply as well as demand of power with electric autos additionally could bring about maintaining a strategic distance from exorbitant moves up to the grid, for example, putting resources into new power plants. In this way, transportation expenses and power bills could reduce with vehicle-to-grid administrations. As more clients embrace EVs, vehicle-to-grid administrations ought to be considered to help level out power supply as well as demand. This alternative might be particularly helpful in urban communities that have embraced electric transports for public transportation. These electric transports could give power to the grid when not being used, diminishing expenses for the city as well as clients. Some alert ought to be taken as more EVs are associated with the grid. Huge spikes popular for power could cause anxiety that could influence soundness, productivity, and working expenses of the grid. Subsequently, the effect of charging an electric vehicle is subject to where it is situated on the grid and the season of day it is charged.

Utilities plan to utilise disseminated assets, for example, sustainable power source generation, storage and demand reaction, to incompletely control charging effects of EVs. Keen grid innovations, for example, progressed metering foundation could demonstrate accommodating in dealing with the charging of EVs. Such gadgets permit charging stations to be incorporated with time-based rates that support off-peak charging. They likewise enable utilities to examine charging station utilisation and charging practices to illuminate speculation choices. Moreover, calculations that successfully plan the charging and releasing of EVs are vital for the grid to work proficiently. Be that as it may, growing such calculations is troublesome because of the irregularity and vulnerability of future occasions. More electric vehicle charging stations in advantageous areas are important to adjust request on the grid and increment accommodation. On the off chance that an electric vehicle needs to charge amid a lengthy, difficult experience trip, it would need to stop at the closest charging station. The nearest charging zone may not be along the driver's way, conceivably expanding power utilisation and diminishing convenience. Microgrids could likewise support unwavering quality while charging EVs in an area or work zone. Small community areas with disseminated assets, for example, solar-power oriented, wind as well as storage would decrease the strain on our power grid. Maybe circling capacity into nearby microgrids would additionally improve power versatility. As clients receive EVs, it is vital to consider all the potential advantages these autos could give to the grid. The grid could turn out to be more adaptable amid peak times for less cost and costly foundation updates could be kept away from with vehicle-to-grid administrations.

3.4. Electric vehicle coordination

Providing supplementary services from EVs is a very plausible option in case the process involves large fleet vehicles [40]. As revealed by Kempton et al. [40] in the year 2001, only one EV can neither unswervingly make it electricity market nor establish transactional engagements with electrical institutions because of the power constraints. The authors, therefore, suggested an aggregator technology, which may serve the purpose of intermediation between the automobiles, the utility organisations as well as the energy market. **Figure 3** vividly demonstrates the aggregator frameworks alongside other forms of the framework [41, 42].

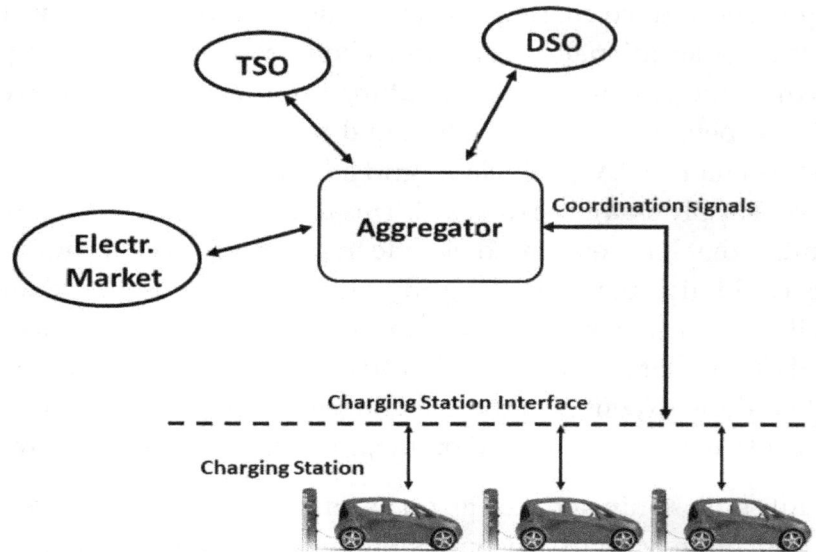

Figure 3. Simplified EV coordination framework: transmission system operator (TSO), distribution system operator (DSO), electric market and charging stations for EVs.

In almost every circumstance, the EV aggregator produces essential signals for coordinating the EV fleet based on the data shared between the supply the energy market, TSO as well as the DSO.

There are multiple reasons behind the deployment of EV aggregators. Firstly, under current market conditions small loads individual participation is prohibited. Also, it permits easiest interaction with the DSO for troubleshooting. With an appropriate strategy, it can lead to a reduction of the risk of forecast errors of the EV load.

As PV penetration increases in LV grid, controlling the EV load can lead to an improvement in the feeder operation and a decrease in the need to invest in infrastructure upgrades of the grid. In grids with such high penetration, there are constraints to keep in mind. Customers will evaluate improvements with respect to dependability, quality, and price. We expect a major change in the near future in the voltage quality improvement, which will aim to reduce long-term variations in voltage magnitude that arise in a decentralised RES generation context. We know from theory, that EV load coordination can facilitate keeping a local balance between production and consumption, which can reduce under and over-voltages.

Planning rules are currently being revised so as to allow for greater RES penetration by European energy suppliers. The options include [43]: (a) redirecting the power route to decrease circuit distance; (b) improvement of the upstream transformer volume; (c) improvement with MV/LV transformer with automatic voltage control; (d) improvement of the grid circuit thermal capacity, (e) improvement of the grid's rated voltage level; (f) setting up of additional reactive power compensation; (g) erection of new substations; and (h) modifications to the grid architecture.

Above are the grid reinforcement options. Recently, automatic power curtailment is being looked at within DSOs, if RES induced voltage variations were to exceed the allowable bounds. To do so, requires linking all units in a communication infrastructure network, which may not be convenient to owners and might imply compensating them for the inconvenience. Extensive research has been made into reactive power methods for LV grids, showing some limitations in the effectiveness of reducing voltage rise [46, 48–50]. In view of the case of PV, most PV units previously connected in LV grids work with unit power factor. Furthermore, the high R/X ratio (imaginary part of impedance divided by the real part of impedance) of LV cables creates challenging the efficiency of these techniques. By providing reactive power contribution by all PV units, there can be visible voltage rise reduction in a feeder [50]. Besides, active power solutions can be utilised as an alternative one.

For the interest of knowledge, Fell et al. [43] explained that an EV aggregator is vital requirement in the coordination of the grid-connected operations of several EVs to achieve the regulation needs or standards by the TSO. It is noteworthy that, the meaning of EV aggregator is also recognised by Brooks and Cage [44]; in this regard, the priority of EV aggregator is to check the driving needs of the operators. Some authors have suggested a strategy that allows the user to communicate their operation needs to the aggregator. Consequently, the aggregator works by processing the information about the driving data. Considering the convenience of all the operation profiles of the user, the aggregator produces a "virtual power plant" in which the number of the automobiles anticipatedly plugged plus the accumulated power as well as the levels of energy. Moreover, the estimation can be done per hour. Equally, Bessa et al. [45] also invented a procedure that can as well be used to optimise and support aggregator in enhancing the engagement of the day-ahead (also known as spot market) as well as subordinate reserves sessions. Further analysis was conducted a 2-year simulation on the Iberian market and concluded that; the agent of aggregation with augmented bidding is likely to reduce the costs of charging more than the coordinated system of billing. Secondly, in case the payment regarding the reserve capacity is convenient, then it is financially expedient to regulate EV participation. Lastly, if there is no reserve capacity compensation, the idea of optimised bidding can as well pay off. Prediction of the EV loads as well as approximating the qualms is essential in problem-solving. Lopes et al. [46] applied the concept y describing the architecture through which EVs are accumulated on a micro-grid as well as multi-grid ideas. Again, the aggregator in this context functions as interplay between the users of EVs and the energy market. In this case, these concepts are like to offer frequency regulation. Researchers Galus and Anderson [47]; introduced a different idea on an aggregator. They asserted that an aggregator is not a corporation but an intangible computational unit that monitors and evaluates a control are: Adequately, it is a smart interface amid the electrical utilities and the EV. Harmonisation of EV is easy to detect if there is the likelihood of responding to the propagated indications. Therefore, the need of EV to react to the secondary services as well as technological shortcomings arising from EV coordination process is addressed by Galus and Anderson [47].

3.4.1. Synergies between EV needs and PV distributed generation

In this subsection, the concept of synergies between undispatchable generation sources and controllable loads, utilising in particular PV distributed generation and EV batteries, is presented. In the smart grid environment, increasing further PV generation, would involve storage systems to modulate power injections. In this situation, EVs (V2G) are more advantageous that conventional energy storage (batteries) within smart grids.

The potential of using RESs as alternative to conventional fuels such as fossil fuels, to power EVs can help reduce pollution with a considerable decarbonisation effect and enhance resource efficiency. Alternatively, sustainability targets are however in advance: more RES volume is expected such as share of renewable energy of 45% in the electricity sector by 2030, which is just currently 21%, and further sustainable transport has to be realised. Consequently, there is a requirement to integrate both elements in the best cost and effective manner.

Figure 4 shows per unit (p.u.) based graphs of the load demand profile (maximum peak load demand) and PV generation (Watt-peak installed PV capacity) against vehicles mobility patterns profile (maximum shares of vehicles in motion. Under unrestricted charging, it is clear that no interdependence can be drawn between the penetration impact of both EV and PV. It can be seen that, that there is no interdependence between the penetration impact of both EV and PV, given that the EV vehicles are charged in unrestricted manner. Actually, it is anticipated that EV generated load peaks should happen during the evening commute home. Besides, the EV demand peak corresponds with periods of little to no energy generation from PV sources, or in best situation to a minimal level throughout the summer when the days are extended.

A greater penetration level of DER along with a mitigation of the intermittent effects of high penetration of PV generation could be facilitated with more flexibility in EV demand. Such synergy requires the transferring of the EV demand peak in order to prevent bottleneck and high electricity prices during peak demand as well to relieve the excess of PV power generation. It is essential to emphasise that the tracked objective is principally focused on risk mitigation instead of minimising operational expenditures. Note that lacking to alleviate risk

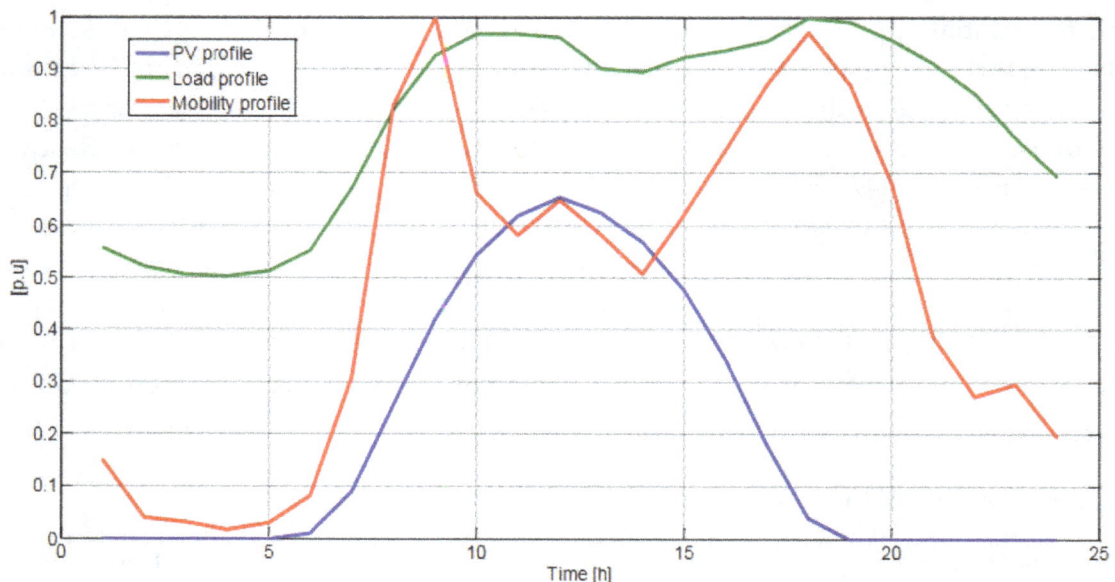

Figure 4. Load demand profile against typical EV load and PV generation profiles as a function of time of day.

could be costly (e.g. forfeit from regulator, energy not delivered or damaging, the standing of the utility company). Smart consumer and smart home comprises funds committed to create the required interfaces and utilities for prosumers interactions and communications with the appropriate objects (i.e. energy aggregators, utilities, DSOs, etc.). This essentially implicates direct funds in terms of white goods; smart meters in addition to the evaluation of prosumers commitment and sensitivity to diverse incentives. Virtual power plants amalgamation is probably the best attractive form regarding the synergies between EV and PV. Its operation is systematically reliant on other divisions.

Such funds concentrate on physical and market aggregation among distributed generation and controllable loads creating jointly a virtual power plan. A virtual power plant assessed to a conventional RES offers numerous benefits principally: greater inertia (aggregated sources), controllability (through storage or demand side management) permitting the option to interact and assist as a market participant whichever by delivering auxiliary amenities or bidding in energy market. The established design and the energy management approach are in agreement with the usual current and anticipated standards in terms of interoperability potential and functionalities. The evaluated EV charging comprises uncontrolled and synchronised charging approaches. The recommended plan intends to empower effective and harmonious communication among the diverse players within the DEMS (decentralised energy management system) domains. For the smart grid and the deregulated energy markets, it is anticipated to get a variety of players exploiting different equipment' suppliers. These performers require to be effectively interconnected and synchronised with interoperable solutions, directing to significant executions costs payable to the sheer number of procedures and involved equipment, functions or applications to be integrated.

4. Conclusion

EVs represent one of the best promising technologies for green and sustainable transportation systems. The high penetration of EVs will have positive effects and benefits such as lesser fossil fuel reliance, significant reduction of GHG and toxic pollutant emissions, as well as the capability to contribute in the integration of renewable energy into existing electric grids. This chapter reviewed the latest advances related to the interaction and integration of EVs with RESs such as wind energy, solar photovoltaics and EV coordination for sustainable mobility in significantly reducing air pollution. Some key concerns and possible solutions were also discussed in detail. The successful implementation of the coupling EV-RES technology includes and requires the full contribution of government, power utilities, EV and aggregators manufactures, policy-makers and owners. It is expected that this study can assist all involved parties to better understand the challenges and issues and contribute further to this field.

Conflict of interest

The authors declare no conflict of interest.

Author details

Michela Longo[1*], Federica Foiadelli[1] and Wahiba Yaïci[2]

*Address all correspondence to: michela.longo@polimi.it

1 Department of Energy, Politecnico di Milano, Milan, Italy

2 CanmetENERGY Research Centre, Natural Resources Canada, Ottawa, Ontario, Canada

References

[1] Environment and Climate Change Canada. Canadian Environmental Sustainability Indicators, Greenhouse Gas Emissions. 2016. Available from: www.ec.gc.ca/indicateurs-indicators/default.asp?lang=en&n=FBF8455E [Accessed: 2018-02-20]

[2] Government of Canada. Canada's Way Forward on Climate Change, The Paris Agreement. 2016. Available from: http://climatechange.gc.ca/default.asp?lang=En&n=24700154-1 [Accessed: 2018-02-20]

[3] Office of the Parliamentary Budget Officer. Canada's Greenhouse Gas Emissions: Developments, Prospects and Reductions. 2016. Available from: http://www.pbo-dpb.gc.ca/web/default/files/Documents/Reports/2016/ClimateChange/PBO_Climate_Change_EN.pdf [Accessed: 2018-02-20]

[4] European Commission. A European Strategy for Low-Mission Mobility. Brussels. 2016. Available from: https://ec.europa.eu/clima/policies/transport_en [Accessed: 2018-03-02]

[5] Thomas CES. Transportation options in a carbon constrained world: Hybrids, plug-in hybrids, biofuels, fuel cell electric vehicles, and battery electric vehicles. International Journal of Hydrogen Energy. 2009;**34**:9279-9296. DOI: 10.1016/j.ijhydene.2009.09.058

[6] Richardson DB. Electric vehicles and the electric grid: A review of modeling approaches, impacts, and renewable energy integration. Renewable and Sustainable Energy Reviews. 2013;**19**:247-254. DOI: 10.1016/j.rser.2012.11.042

[7] Mwasilu F, Justo JJ, Kim E-K, Do TD, Jung J-W. Electric vehicles and smart grid interaction: A review on vehicle to grid and renewable energy sources integration. Renewable and Sustainable Energy Reviews. 2014;**34**:501-516. DOI: 10.1016/j.rser.2014.03.031

[8] Liu L, Kong F, Liu X, Peng Y, Wang Q. A review on electric vehicles interacting with renewable energy in smart grid. Renewable and Sustainable Energy Reviews. 2015;**51**:648-661. DOI: 10.1016/j.rser.2015.06.036

[9] European Commission. Community Research and Development and Development Information Service (CORDIS), eMobilita, Electromobility in urban transport: A multi-dimensional innovation (socio-economic and environmental effects). Available from: https://cordis.europa.eu/project/rcn/207049_en.html [Accessed: 2018-03-25]

[10] European Commission, Community Research and Development and Development Information Service (CORDIS), NeMo: Hyper-Network for electroMobility. Available from: https://cordis.europa.eu/project/rcn/204973_en.html [Accessed: 2018-03-25]

[11] European Commission, Community Research and Development and Development Information Service (CORDIS), OSEM-EV (Optimised and Systematic Energy Management in Electric Vehicles). Available from: http://cordis.europa.eu/project/rcn/194883_en.html [Accessed: 2018-03-25]

[12] European Environment Agency. Electric vehicles and the energy sector—Impacts on Europe's future emissions. 2016. Available from: https://www.eea.europa.eu/themes/transport/electric-vehicles/electric-vehicles-and-energy [Accessed: 2018-02-20]

[13] Tan KM, Ramachandaramurthy VK, Yong JY. Integration of electric vehicles in smart grid: A review on vehicle to grid technologies and optimization techniques. Renewable and Sustainable Energy Reviews. 2016;**53**:720-732. DOI: 10.1016/j.rser.2015.09.012

[14] Dubarry M, Devie A, McKenzie K. Durability and reliability of electric vehicle batteries under electric utility grid operations: Bidirectional charging impact analysis. Journal of Power Sources. 2017;**358**:39-49. DOI: 10.1016/j.jpowsour.2017.05.015

[15] Shaukata N, Khan B, Ali SM, Mehmood CA, Khan J, Farid U, Majid M, Anwar SM, Jawad M, Ullah Z. A survey on electric vehicle transportation within smart grid system. Renewable and Sustainable Energy Reviews. 2018;**81**:1329-1349. DOI: 10.1016/j.rser.2017.05.092

[16] Pinson P, Christensen LEA, Madsen H, Sørensen PE, Donovan MH, Jensen LE. Regime-switching modelling of the fluctuations of offshore wind generation. Journal of Wind Engineering & Industrial Aerodynamics. 2008;**96**(2):2327-2347. DOI: 10.1016/j.jweia.2008.03.010

[17] Jabr RA, Pal BC. Intermittent wind generation in optimal power flow dispatching. IET Generation, Transmission & Distribution. 2009;**3**(1):66-74. DOI: 10.1049/iet-gtd:20080273

[18] Bollen MHJ, Zhong J, Zavoda F, Meyer J, McEachern A, Corcoles Lopez F. Power quality aspects of smart grids. In: Proceedings of International Conference on Renewable Energies and Power Quality (ICREPQ'10); 23-25 March 2010; Granada, Spain. DOI: 10.24084/repqj08.58

[19] Kempton W, Letendre SE. Electric vehicles as a new power source for electric utilities. Transportation Research Part D: Transport and Environment. 1997;**2**(3):157-175. DOI: 10.1016/S1361-9209(97)00001-1

[20] Clement-Nyns K, Haesen E, Driesen J. The impact of charging plug-in hybrid electric vehicles on a residential distribution grid. IEEE Transactions on Power Systems. 2010;**25**(1):371-380. DOI: 10.1109/TPWRS.2009.2036481

[21] Kempton W, Tomic J. Vehicle-to-grid power implementation: From stabilizing the grid to supporting large-scale renewable energy. Journal of Power Sources. 2005;**144**(1):280-294. DOI: 10.1016/j.jpowsour.2004.12.022

[22] Kempton W, Udo V, Huber K, Komara K, Letendre S, Brunner SD, Pearre N. A test of vehicle-to-grid (V2G) for energy storage and frequency regulation in the PJM system. Technical Report, MAGIC Consortium, Jan., 2009. Available at: http://www.magicconsortium.org/_Media/test-v2g-in-pjm-jan09.pdf [Accessed: 2018-02-20]

[23] Pillai JR, Bak-Jensen B. Integration of vehicle-to-grid in the western Danish power system. IEEE Transactions on Sustainable Energy. 2011;2(1):12-19. DOI: 10.1109/TSTE.2010. 2072938

[24] Galus MD, Koch S, Andersson G. Provision of load frequency control by PHEVs, controllable loads, and a cogeneration unit. IEEE Transactions on Industrial Electronics. 2011;58(10):4568-4582. DOI: 10.1109/TIE.2011.2107715

[25] Hay C, Togeby M, Bang NC, Søndergren C, Hansen LH. Introducing electric vehicles into the current electricity markets. Project Report, EDISON Deliverable D2.3. May 2010

[26] Energinet.dk. Energinet.dk's ancillary services strategy. Aug. 2011. Available from: www. energinet.dk [Accessed: 2018-02-20]

[27] Lopes JAP, Almeida PMR, Soares FJ. Using vehicle-to-grid to maximize the integration of intermittent renewable energy resources in islanded electric grids. In: Proceedings of the International Conference on Clean Electrical Power (ICCEP '09); 9-11 June 2009; Capri, Italy. DOI: 10.1109/ICCEP.2009.5212041

[28] Lopes JAP, Polenz SA, Moreira CL, Cherkaoui R. Identification of control and management strategies for LV unbalanced microgrids with plugged-in electric vehicles. Electric Power Systems Research. 2010;80(1):898-906. DOI: 10.1016/j.epsr.2009.12.013

[29] Bessa RJ, Matos MA. Economic and technical management of an aggregation agent for electric vehicles: A literature survey. International Transactions on Electrical Energy Systems. 2011;22:334-350. DOI: 10.1002/etep.565

[30] Birnie D. Solar-to-vehicle (S2V) systems for powering commuters of the future. Journal of Power Sources. 2009;186(2):539-542. DOI: 10.1016/j.jpowsour.2008.09.118

[31] Bollen M, Hassan F. Integration of distributed generation in the power system. Chapter 5: Voltage Magnitude Variations. 5th ed. John Wiley & Sons Inc.; 2011. DOI: 10.1002/ 9781118029039

[32] Corfee K, Korinek D, Cassel W, Hewicker C, Zillmer J, Pereira Morgado M, Ziegler H, Tong N, Hawkins D, Cernadas J. Distributed generation in Europe–Physical infrastructure and distributed generation connection, Memo. Available from: http://www. clean-coalition.org/site/wp-content/uploads/2012/11/Memo-1_Physical-Infrastructure-and-DG-Interconnection.pdf [Accessed: 2018-02-20]

[33] Etherden N, Bollen MHJ. Increasing the hosting capacity of distribution networks by curtailment of renewable energy sources. In: Proceedings of PowerTech; 19-23 June 2011; Trondheim, Norway. DOI: 10.1109/PTC.2011.6019292

[34] Carvalho P, Correia P, Ferreira L. Distributed reactive power generation control for voltage rise mitigation in distribution networks. EEE Transactions on Power Systems. 2008;**23**(2):766-772. DOI: 10.1109/TPWRS.2008.919203

[35] Tonkoski R, Lopes LAC, El-Fowly T. Coordinated active power curtailment of grid connected PV inverters for overvoltage prevention. IEEE Transactions on Sustainable Energy. 2011;**2**(2):139-147. DOI: 10.1109/TSTE.2010.2098483

[36] Fawzy T, Premm D, Bletterie B, Gorsek A. Active contribution of PV inverters to voltage control—From a smart grid vision to full-scale implementation. e & i Elektrotechnik und Informationstechnik. 2011;**128**(4):110-115. DOI: 10.1007/s00502-011-0820-z

[37] Demirok E, Casado Gonzalez P, Frederiksen KHB, Sera D, Rodriguez P, Teodorescu R. Local reactive power control methods for overvoltage prevention of distributed solar inverters in low-voltage grids. IEEE Journal of Photovoltaics. 2011;**1**(2):174-182. DOI: 10.1109/JPHOTOV.2011.2174821

[38] Marra F, Fawzy YT, Bülo T, BlažicBlazic B. Energy storage options for voltage support in low-voltage grids with high penetration of photovoltaic. In: Proceeding 2012 3rd IEEE PES International Conference and Exhibition on Innovative Smart Grid Technologies (ISGT Europe '13); 14-17 October 2012; Berlin, Germany. DOI: 10.1109/ISGTEurope.2012.6465690

[39] Liu X, Aichhorn A, Liu L, Li H. Coordinated control of distributed energy storage system with tap changer transformers for voltage rise mitigation under high photovoltaic penetration. IEEE Transactions on Smart Grid. 2012;**3**(2):897-906. DOI: 10.1109/TSG.2011.2177501

[40] Kempton W, Tomic J, Letendre SE, Brooks A, Lipman T. Vehicle to grid power: Battery, hybrid, and fuel cell vehicles as resources for distributed electric power in California. Working Paper Series ECD-ITS-RR-01-03, Jun. 2001. Available from: http://escholarship.org/uc/item/5cc9g0jp [Accessed: 2018-02-20]

[41] Commission of the European Communitie. Regulation (EEC) No 4064/89 Merger Procedure. Mar. 1999. [Online]. Available from: www.ec.europa.eu [Accessed: 2018-02-20]

[42] Galus MD, Zima M, Andersson G. On integration of plug-in hybrid electric vehicles into existing power system structures. Energy Policy. 2010;**38**(11):6736-6745. DOI: 10.1016/j.enpol.2010.06.043

[43] Fell K, Huber K, Zink B. Assessment of plug-in electric vehicle integration with ISO/RTO systems. Technical Report, Mar. 2010. Available from: http://www.isorto.org/atf/cf/%7B5B4E85C6-7EAC-40A0-8DC3003829518EBD%7D/IRC_Report_Assessment_of_Plugin_Electric_Vehicle_Integration_with_ISORTO_Systems_03232010.pdf [Accessed: 2018-02-20]

[44] Brooks A, Gage T. Integration of electric drive vehicles with the electric power grid—A new value stream. In: 18th Internat. Electric Vehicle Symposium and Exhibition, Oct. 2001

[45] Bessa RJ, Matos MA, Soares FJ, Lopes JAO. Optimized bidding of a EV aggregation agent in the electricity market. IEEE Transactions on Smart Grid. 2012;**3**(1):443-452. DOI: 10.1109/TSG.2011.2159632

[46] Lopes JAP, Soares FJ, Almeida PMR. Integration of electric vehicles in the electric power system. In: Proceeding of the IEEE, 04 October 2010; pp. 168-183. DOI: 10.1109/JPROC.2010.2066250

[47] Galus MD, Andersson G. Demand management of grid connected plug-in hybrid electric vehicles. In: Proceedings of the Energy 2030 Conference (ENERGY 2008); 17-18 November 2008; Atlanta, GA, USA. DOI: 10.1109/ENERGY.2008.4781014

[48] Brenna M, Foiadelli F, Longo M. The exploitation of vehicle-to-grid function for power quality improvement in a smart grid. IEEE Transactions on Intelligent Transportation Systems. 2014;**15**(5):2169-2177. DOI: 10.1109/TITS.2014.2312206

[49] Kanchev H, Di Lu F, Colas V, Lazarov B. Energy management and operational planning of a microgrid with a PV-based active generator for smart grid applications. IEEE Transactions on Industrial Electronics. 2011;**58**(10):4583-4592. DOI: 10.1109/TIE.2011.2119451

[50] Liserre M, Sauter T, Hung JY. Future energy systems, integrating renewable energy sources into the smart power grid through industrial electronics. IEEE Industrial Electronics Magazine. 2010;**4**(1):18-37. DOI: 10.1109/MIE.2010.935861

5

Model-Based System Design for Electric Vehicle Conversion

Ananchai Ukaew

Abstract

Development of electric vehicle (EV) conversion process can be implemented in a low-cost and time-saving manner, along with the design of actual components. Model-based system design is employed to systematically compute the power flow of the electric vehicle propulsion and dynamic load. Vehicle specification and driving cycles were the two main inputs for the simulation. As a result, the approach is capable of predicting various EV characteristics and design parameters, such as EV performance, driving range, torque speed characteristics, motor power, and battery power charge/discharge, which are the necessity for the design and sizing selection of the main EV components. Furthermore, drive-by-wire (DBW) ECU function can be employed by means of model-based design to improve drivability. For the current setup, the system components are consisted of actual ECU hardware, electric vehicle models, and control area network (CAN) communication. The EV component and system models are virtually simulated simultaneously in real time. Thus, the EV functionalities are verified corresponding to objective requirements. The current methodology can be employed as rapid design tool for ECU and software development. Same methodology can be illustrated to be used for EV tuning and reliability model test in the future.

Keywords: EV conversion, model-based system design, drive-by-wire ECU, real-time application, in-the-loop testing, rapid control design, ECU network, CAN protocol

1. Introduction

Development in EV conversion has been vastly improved in the recent year. However, different vehicle models have different technical specifications, so conversion kits for each one of them have to be customized in order to meet the specific requirement such as range per charge and acceleration performance. Engineers, therefore, have to make the decision on the capacity of batteries and also how many of them are required to meet the driving demand. Moreover, selection of different types of motor is also presented as the main requirement [1]. Normal design process would require high-end expensive software to model the EV system. Furthermore, building the EV without the knowledge of the parameters within the system could costly lead to the failure of the design.

In addition, poor vehicle performance safety and reliability might occur when new electric propulsion characteristics do not match with the characteristics of replaced engine sharing the same chassis.

Therefore, a sub-ECU must be developed to harmonize EV propulsion dynamics and existing vehicle chassis characteristics called drive-by-wire (DBW) [2]. DBW functionality can then improve EV drivability by providing power demand to the electric motor drive according to the driver preference. However, installation of the DBW ECU without appropriate functional safety design and evaluation could induce such system failures or component malfunctions due to unpredicted behaviors during actual driving situations. Therefore, during the initial development process, ECU functions are needed to be established and evaluated against design and functional safety aspects beforehand [3, 4].

To improve EV conversion development process, model-based design process is shown in **Figure 1**. The method would benefit the design engineer in making better decision for the conversion and also saving time and cost by reducing error during the design process [5–7]. The process can be employed to perform system simulation based on different scenarios and technical specification. Embedded system and DBW ECU can be realized by software rapid auto coding to shorten error correction and debugging time. Virtual prototyping test can be employed to validate design requirement and EV conversion specification. The in-the-loop tests can ensure accurate implementation of both software and hardware ECU for the conversion using real-time verification methodology.

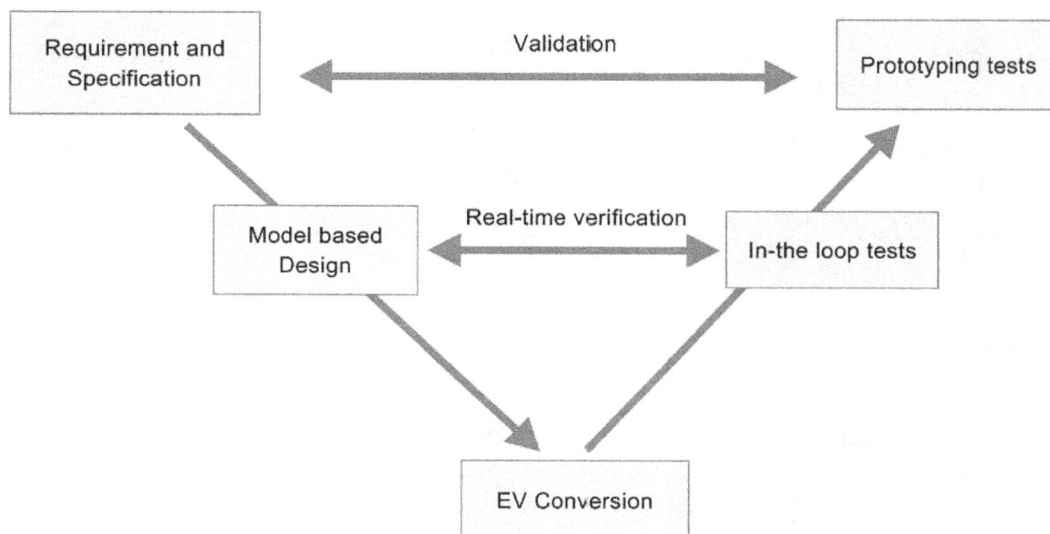

Figure 1. Model-based design process for EV conversion.

In this literature, the first model-based design for EV conversion prototyping development, which describes electric vehicle modeling including EV traction, EV components, and power flow models, is defined. Then, drive-by-wire ECU design and in-the-loop testing for EV conversion process are described in details. The last section illustrates versatility of model-based design in EV conversion tuning and diagnostic application.

2. EV conversion prototyping development

2.1. EV system modeling

In order to set up the simulation of EV, mathematical models have to be generated first from the engineering principles and theories. The four core models are traction model, motor model, battery model, and power flow model as follows.

2.1.1. Traction model

Forces acting on the vehicle govern the equation for vehicle traction as seen in **Figure 2**. Those forces comprised of tractive forces (F_{te}), rolling resistance force (F_{rr}), aerodynamic force (F_{ad}), lateral acceleration force (F_{la}), wheel acceleration force (F_{wa}), hill climbing force (F_{hc}) [or component force of vehicle weight which depend on grade (θ)], and the gross weight of the co EV (mg).

The governing relation can be found in Eq. (1) where traction needs to overcome the load that is equal to five other forces:

$$F_{te} = F_{rr} + F_{ad} + F_{hc} + F_{la} + F_{wa} \tag{1}$$

where equation for each force components can be employed from many sources such as reference [2, 4] and other automotive textbooks.

2.1.2. Motor efficiency model

In the EV conversion system, the motor replaces the internal combustion engine (ICE) in providing the torque to drive the wheel as shown in **Figure 3**, which also affects the traction

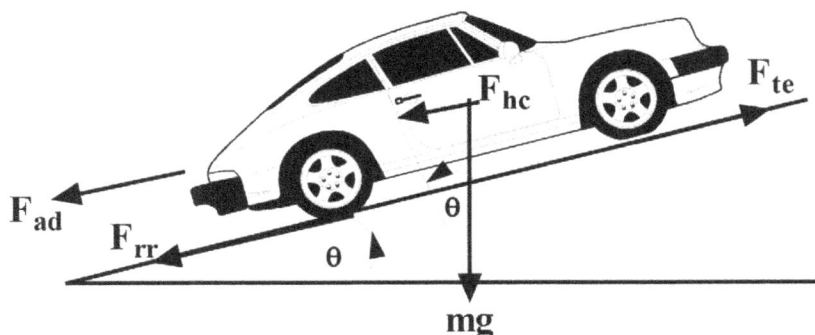

Figure 2. The force components involved in the vehicle traction.

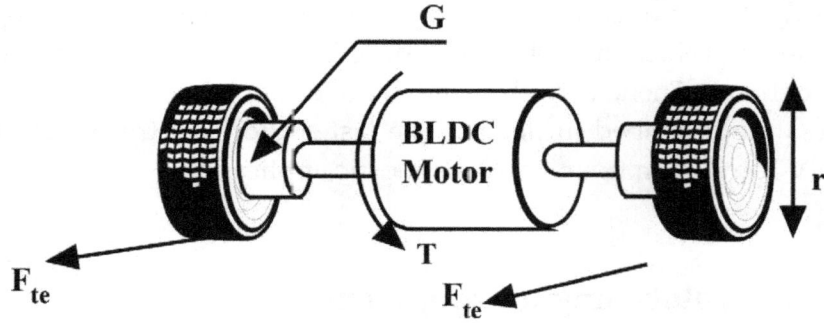

Figure 3. The EV motor provides the traction to the vehicle through transmission.

of the vehicle. The motor torque, speed, and efficiency equation are presented in Eqs. (2), (3), and (4), respectively.

- Motor torque (T)

$$T = \frac{F_{te}r}{G} \quad \text{N.m} \tag{2}$$

- Motor angular speed (ω)

$$\omega = G\frac{v}{r} \quad \text{rad.s}^{-1} \tag{3}$$

- Motor efficiency (η_m)

$$\eta_m = \frac{T\omega}{T\omega + k_c T^2 k_i \omega + k_\omega \omega^3 + C} \tag{4}$$

where k_c is copper losses coefficient, k_i is iron losses coefficient, k_ω is windage loss coefficient, and C is constant loss applied at any speed [6].

2.1.3. Battery discharge model

Battery's dynamic behavior does have a great effect on EV performance and range. Three common types of batteries, which are lead acid, nickel cadmium, and lithium ion batteries, are governed by Eqs. (5), (6), and (7) here, respectively. As seen in [4, 6], open-circuit voltage (E) of the batteries is changed as the state of charge changes and is calculated for each battery type below:

- Lead acid:

$$E = n \cdot \left(2.15 - DoD \cdot [2.15 - 2.00]\right) \tag{5}$$

- Nickel cadmium:

$$E = n \cdot \left(\begin{array}{l} -8.2816DoD^7 + 23.5749DoD^6 - 23.7053DoD^4 - 12.5877DoD^3 + \\ 4.1315DoD^2 - 8.65DoD + 1.37 \end{array} \right) \tag{6}$$

- Li-ion:

$$E = n \cdot 3.3 \tag{7}$$

(nominal cell voltage = 3.3 V up to 80% DoD)

where n is the number of cells and DoD is depth of discharge (0–1).

The open-circuit voltage also affects the battery current (I_B) in both states of charge and discharge as seen in Eqs. (8) and (9).

- Battery current discharge operating at power (P_{bat})

$$I_B = \frac{E - \sqrt{E^2 - 4RP_{bat}}}{2R} \tag{8}$$

- Battery current charge during regenerative braking

$$I_B = \frac{-E + \sqrt{E^2 + 4RP_{bat}}}{2R} \tag{9}$$

where R is the battery resistance. Due to Peukert phenomenon [6], therefore it is necessary to take into account such effect by adding the power to the k value, such as lead acid battery ($k \approx 1.12$) and Lithium ion ($k \approx 1$), when simulation of battery discharge is performed. Battery capacity (CR) is updated for each time step (δt) as shown in Eq. (10) and then used to update the depth of discharge (DoD) in Eq. (11) for discharging state and in Eq. (12) for charging state as following:

- Total charge removed from battery by the nth step of the simulation

$$CR_{n+1} = CR_n + \frac{\delta t \cdot I_B^k}{3600} \tag{10}$$

- The depth of discharged

$$DoD_n = \frac{CR_n}{C_p} \tag{11}$$

- Charge removed for regenerative braking

$$CR_{n+1} = CR_n - \frac{\delta t \cdot I_B}{3600} \tag{12}$$

2.1.4. EV conversion system power flow model

To complete the simulation, the integrated power flow model is necessary to compute and update the rate of energy going in and out of battery cells, accessories, the motor, gearing components, and wheel to the road and back. Therefore, the model needs to be capable of mathematically simulating the power flow in both driving and braking as shown in **Figure 4**.

Traction model provides the power flow between the vehicle and the road (P_{te}) as shown in Eq. (13). Furthermore, the motor model provides the power going in for both driving and braking mode at the motor/battery connection (P_{mot_in}) and at the motor/transmission connection (P_{mot_out}) as indicated in Eqs. (14) and (15). The power parameters are affected by the motor efficiency (η_m) and the gearing efficiency (η_g). The battery power is also computed and updated (Eq. (16)) during charge and discharge operation using the battery model. Power (P_{ac}) is constantly drawn out of battery due to the use of accessories, such as car stereo and light, which is accounted in Eq. (16) [6, 8].

- Energy required per second:

$$P_{te} = F_{te} \cdot v \tag{13}$$

- Motor power in driving mode:

$$P_{mot\_in} = \frac{P_{mot\_out}}{\eta_m},$$
$$P_{mot\_out} = \frac{P_{te}}{\eta_g} \tag{14}$$

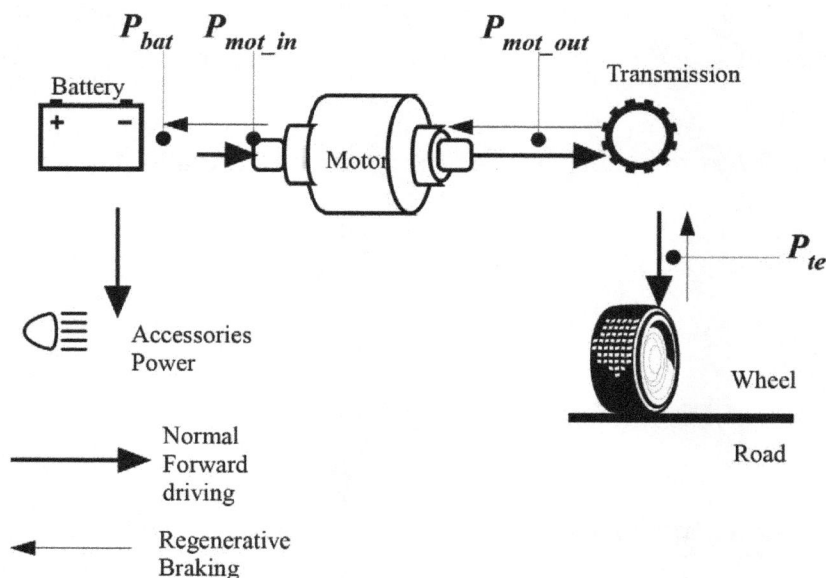

Figure 4. Diagram show power flow in/out components within the EV system for both normal forward driving and regenerative braking operations [6].

- Motor power in braking mode:

$$P_{mot\_in} = P_{mot\_out} \cdot \eta_m,$$
$$P_{mot\_out} = P_{te} \cdot \eta_g$$

(15)

- Battery power:

$$P_{bat} = P_{mot\_in} + P_{ac}$$

(16)

2.2. EV conversion simulation test

Models described in the previous section, especially traction model, are employed to simulate the electric vehicle conversion (EVC) performance by obtaining the velocity plot. The vehicle model specifications are approximately used as the input for the simulation. Other inputs are motor specification and road condition where Refs. [2, 4, 8] explained this specification in details.

2.2.1. Programming for simulation

The traction model is reduced to nonlinear first-order differential forms in [6, 8] when all inputs are substituted. Then, differential equation of velocity is numerically solved using the MATLAB script (.m) file for each time step and updates the values in the program arrays. The out velocity can be plotted against time. The EVC performance here is specified as the time for vehicle to accelerate from 0 to 100 km/h.

2.2.2. EV driving simulation

The other important piece of information for the EV design is range per charge, which tells us how far the vehicle can travel before it needs to be recharged again. In order to obtain such information, the motor model, battery model, and power flow model introduced in the previous section are applied here along with additional inputs. Driving cycle needs to be reasonably selected to simulate the driving dynamics. For present simulation, simplified federal urban driving cycle (SFUD) in [6] is chosen since the vehicle is expected to be driven in the urban area most of the time. The main program [6, 8] is employed to call inputs, including vehicle specification and driving cycle, and then execute the power flow model and battery model for each driving cycle and update parameters, such as range and *DoD* simultaneously. The range per charge then can be plotted when the program is done executing the program. Scenarios for EV range design can be explored using this simulation procedure [6, 8].

2.2.3. EV conversion design parameter simulation

EV design parameters shown in the list below can be easily obtained using the simulation done earlier. To obtain such information, we need to simply write the MATLAB commands in EV main program to update our interested parameters and then write the plot command.

An example of such torque speed map plots is shown in **Figure 5**, and the vehicle is still operated within the motor power range and maximum torque of 250 Nm. The constant torque

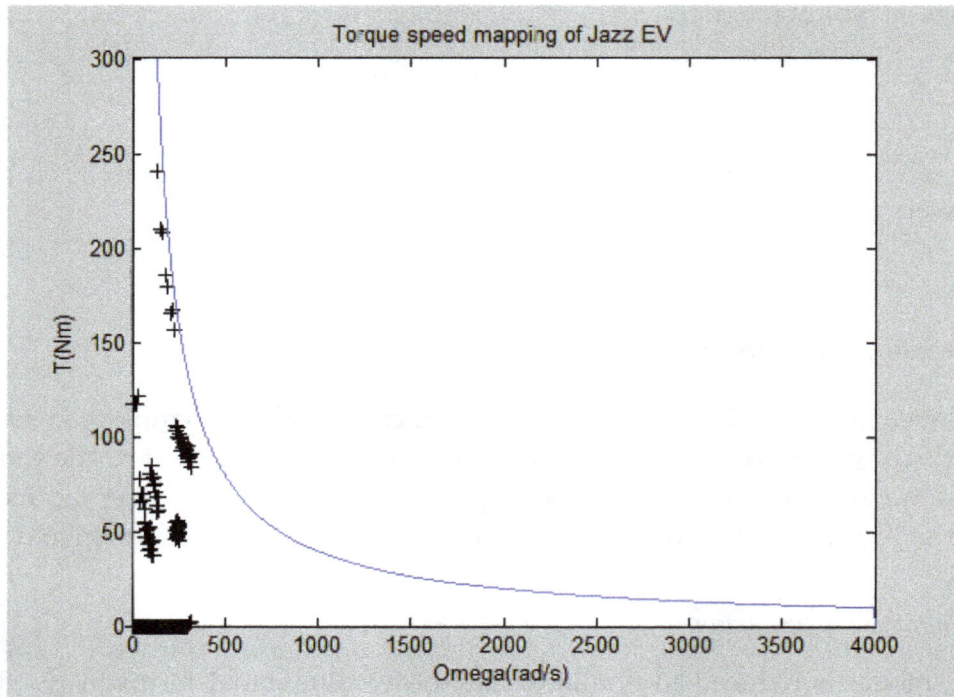

Figure 5. Torque speed map of EVC with SFUD driving cycle and no regenerative braking mode.

region is quite small compared to the field weakening region. The plot also reveals that low motor speed is mostly required when driving in the urban area.

3. EV conversion ECU design and in-the-loop testing

In current EV conversion development as shown in **Figure 6**, drive-by-wire (DBW) functions were developed by means of model-based design approach to synchronize the EV driving characteristics and to improve its drivability. The process starts by setting up parameters and variables for conceptual ECU system requirements. Then, the ECU I/O and signals for communications are formulated. Here, both software functions and embedded hardware design for DBW ECU should be completely determined. The next process is to virtually test DBW ECU against requirements' virtually simulated environment. In this process, the main functionalities along with faulty software or malfunction situation for the ECU can be tested. Bug in the software or communication can be tested and tuned safely with this in-the-loop testing methodology throughout the development process.

3.1. Drive-by-wire ECU design

The main function of conceptual drive-by-wire ECU developed by [2] is to determine power demand from the driver, through vehicle supervisory control ECU, based on the pedal ratio in percentage as shown in **Figure 7** and software algorithm. Next, percent pedal kickdown signal is sent to DWW ECU for torque command and regenerative percentage setting based on power

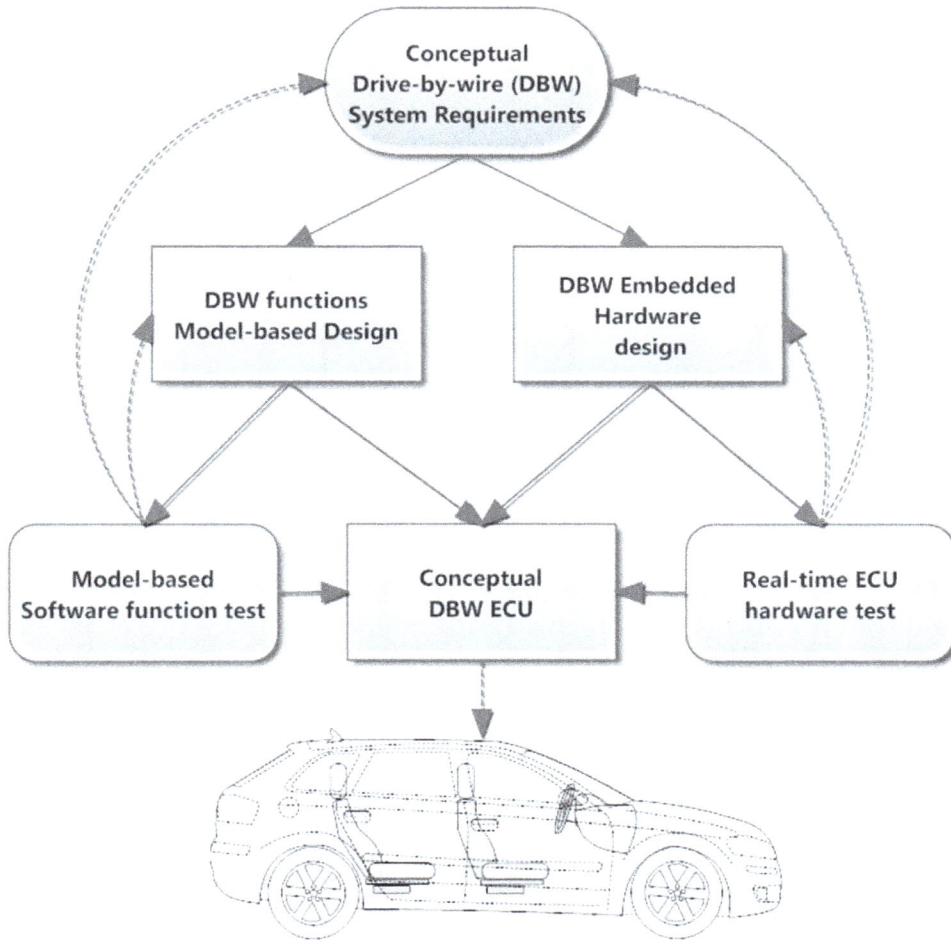

Figure 6. Novel methodology for rapid and safe EV software and hardware development [9].

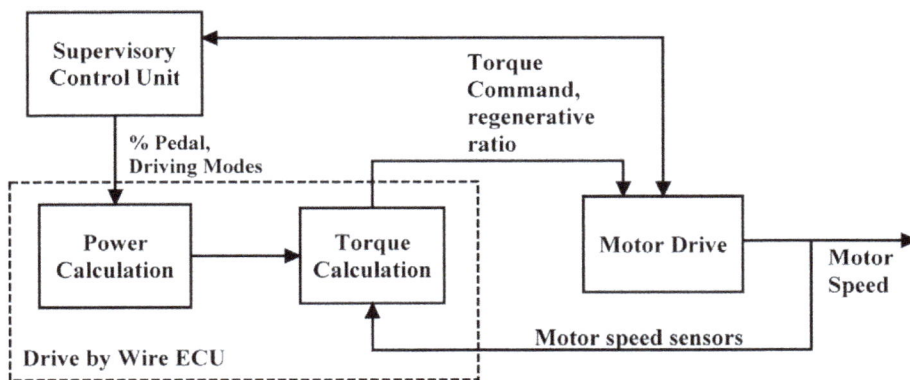

Figure 7. Drive-by-wire ECU functions and signal connection to the supervisory ECU and the motor drive unit [2].

to torque calculation and motor speed signals [9] in rule-based control algorithm. The input and output (I/O) parameters employed for DBW software and ECU are presented in **Figure 8**.

Since new characteristics from EV propulsion are applied to the old chassis. New torque map shall be calculated to compensate EV conversion performance. The design process can be

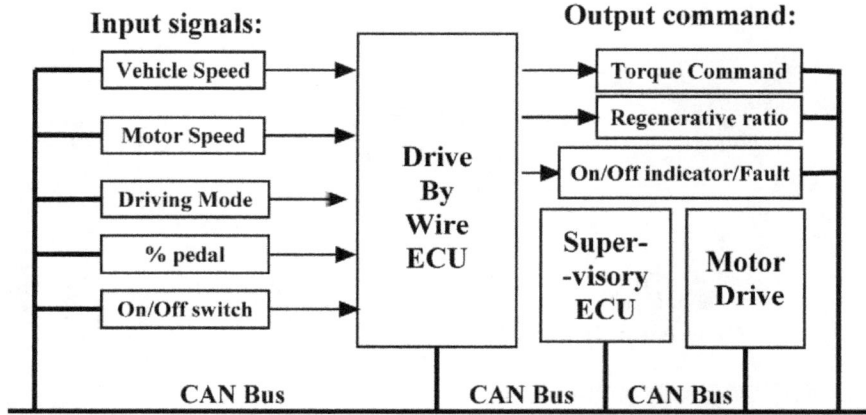

Figure 8. Input and output signal flow of the drive-by-wire ECU with CAN bus interface [2].

reviewed in Ref. [2]. The basic principle is to determine torque setting for various EV driving situations in four quadrants of torque speed map. This methodology can enable the design of more advanced features, such as driving modes, and other advanced driver assistance system (ADAS).

3.2. Model-in-the-loop test

Initial concept of EV software functions can be tested by simulating EV system components and virtual environment of model-based software function test as seen in **Figure 9** [4, 10, 11]. Model-based system design of EV component and drive-by-wire algorithm development can be consulted in details in Ref. [2]. In-the-loop models of driving test profile, supervisory control, DBW function, and motor are developed by using simulation software such as MATLAB/Simulink or others to emulate EV parameters and communication between the ECU and the driving load from vehicle dynamic model. System design requirement can be verified in this MBSF testing stage, such as toque map, and driving mode tests, which can be

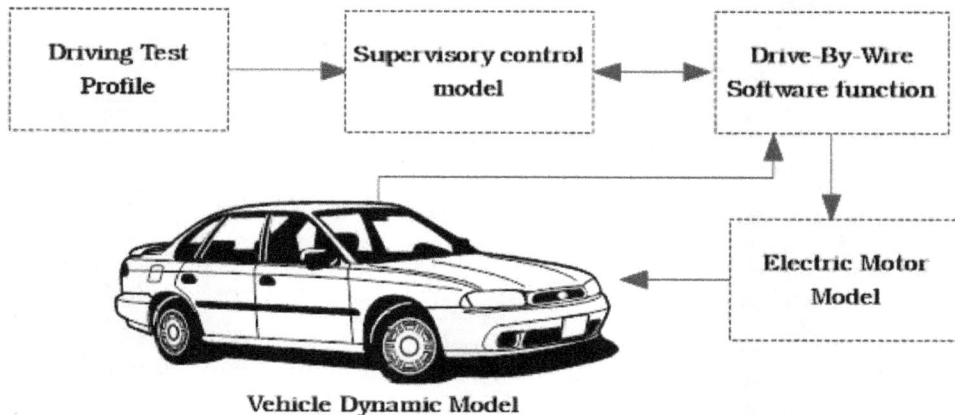

Figure 9. Model-based software function test setup where drive-by-wire function model is connected within the loop with other joint models [9].

done by setting up driving test profile and run the simulation for analysis. However, its capability is not as effective as real-time simulation, which is presented in Section 3.4.

3.3. Simulation analysis

Simulation results from model-in-the-loop test can be analyzed to verify whether system requirements are met. Driving profile in **Figure 14** can be set in several driving schedules as seen in **Figure 10**. After simulation is performed for DBW function, the parameters such as torque speed curve can be analyzed to check EV output such as performance in driving quadrants in **Figure 10**.

Without actual driving, DBW parameters resulted from simulation can be analyzed in different scenarios such as forward driving and reverse driving. Major error can be corrected at this stage along with fault-tolerant function test such as limp home mode in case the DBW is disconnected or malfunctioned.

3.4. Hardware-in-the-loop test

When ECU hardware is ready for testing, software can embed into the ECU to operate in real-time environment. The process is called hardware-in-the-loop (HIL) test when the drive-by-wire software algorithm is replaced by a physical ECU hardware while still connected to virtual environment as seen in **Figure 11**. Thus, HIL components consisted of an actual hardware, real-time interface, and virtual environments (models). It requires a capable communication protocol to handle real-time signal process where CAN protocol is chosen for this HIL System [12–17]. The overall specification of HIL system can be found in **Table 1**.

Arrangement of HIL configuration allows the engineers to conduct test for DBW ECU where it is difficult to perform with the actual vehicle. EV fault and malfunction scenarios can be simulated within the system to check ECU resiliency and fault-tolerant setting. Test repetition and automation can simply be done by scheduling HIL system. Therefore, it helps to reduce testing time and test cases required for real driving. ECU performance testing can be

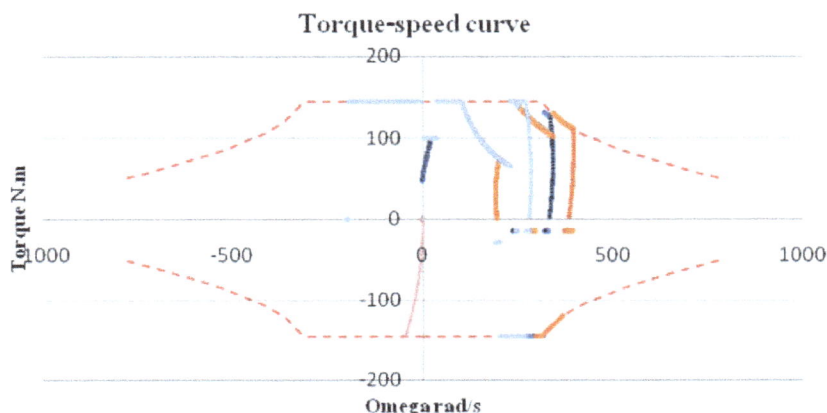

Figure 10. EV torque speed in four-quadrant driving results for analysis by means of model-in-the-loop simulation [8].

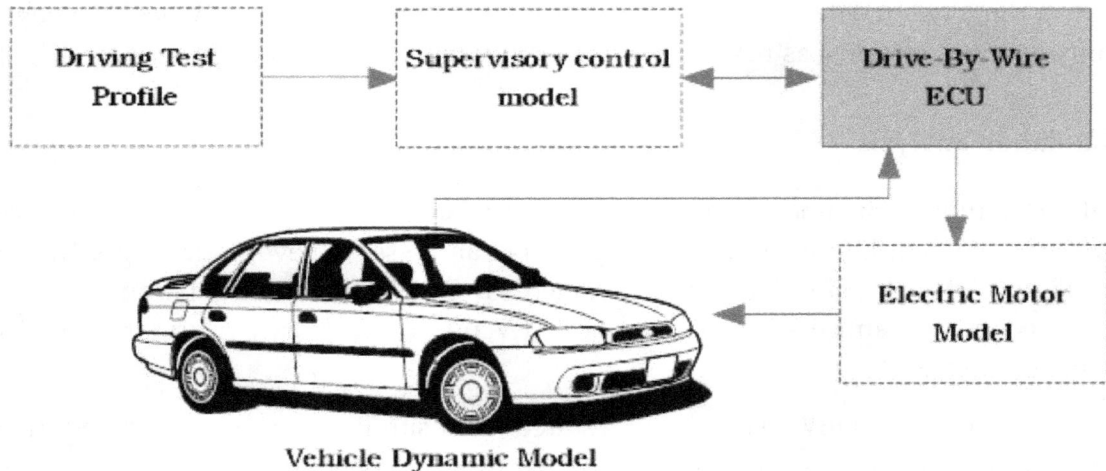

Figure 11. Hardware-in-the-loop (HIL) test configuration for drive-by-wire (DBW) ECU [9].

Components	Specification
Drive-by-wire ECU	Real-time rapid prototyping board
	CPU: ARM Cortex-M4 32bits 168 MHz
	RAM: 8 Mb
Vehicle dynamic and driving profile real-time applications	Real-time processor board
	CPU: ARM Cortex-M4 32bits 168 MHz
	RAM: 8 Mb
Real-time platform	MathWork Simulink real-time workshop
Interface	CAN bus 2.0 (high speed)
	Baud rate: 500 kBaud
Physical connection	CAN: DB9 connector
Power supply	12 V terminal
Protocol sampling time	10 ms

Table 1. Details of HIL test system specification [9].

conducted for EV high speed where it is difficult for real driving test. All model and ECU parameters can be adjusted simultaneously during the test, in real time, enabling more accurate parameter tuning. Therefore, system requirements can be verified in real time in this process.

3.5. Real-time ECU test analysis

To perform DBW ECU HIL test for this work, Simulink real-time workshop toolbox is chosen along with real-time application module for driving profile and vehicle dynamics, and CAN

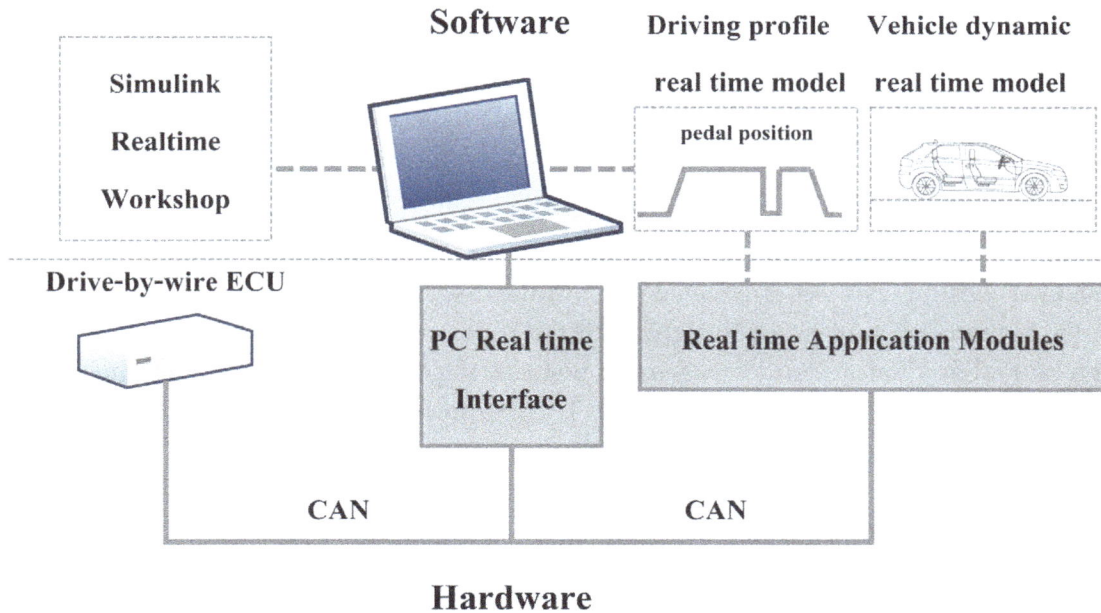

Figure 12. Hardware component integration setup for ECU function tests [9].

protocol is set for PC and ECU real-time interface as seen in **Figure 12**. Simple driving profile for this specific DBW test consists of different driving patterns to represent accelerator pressing by the driver in **Figure 13**. More complicated driving profile can be designated based on test scenarios and particular interest.

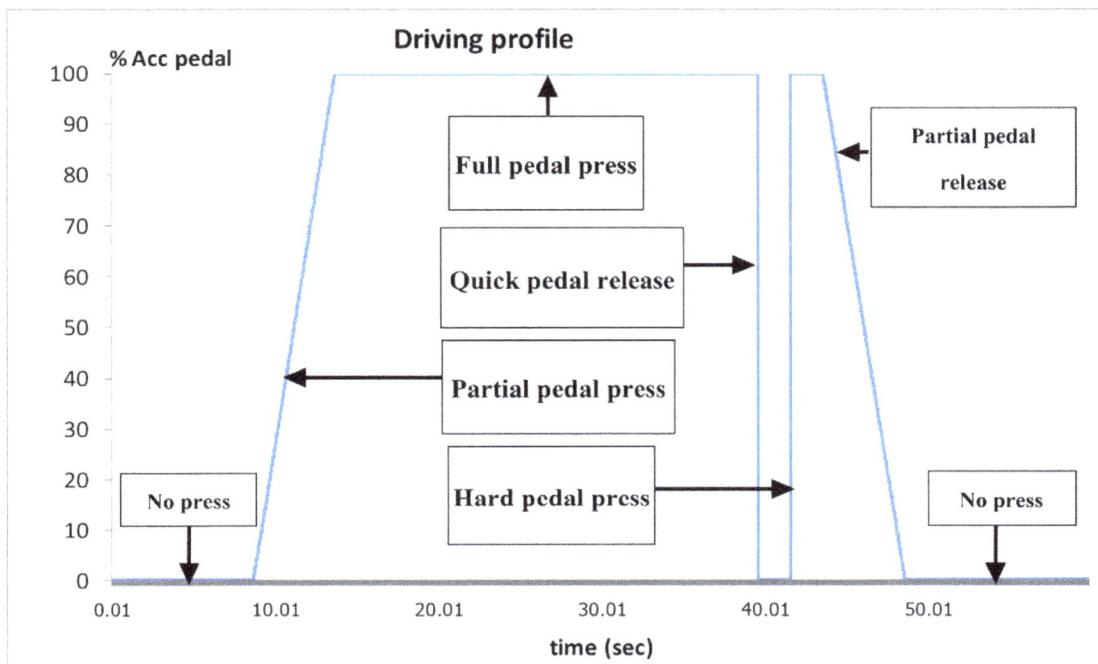

Figure 13. Driving profile test profile based on accelerator pedal position [9].

Based on the driving profile, the test results can be analyzed. In-depth test analysis can be consulted in Ref. [9]. In this case, performance parameters such as vehicle speed and acceleration are monitored for ECU validation.

4. EV conversion tuning and diagnostics

Model-based design can be employed to simulate scenarios where problem might occur during EV conversion process or to figure out the root cause of the problem [18]. The problem can then be realized beforehand by scenario test run to prevent the damage of the EV components and saving time to reinstall new part. The examples of model-based design for tuning and diagnostics can be demonstrated below.

4.1. Motor sizing mismatch scenario

This problem can occur during test run of EV conversion prototype where the motor drive converter becomes too hot during initial run of the EV. This problem forces the drive component to shut down to prevent circuit overheat. Therefore, the EV has to stop early to be checked. The torque map simulation can be executed to check whether there is mismatch between driving load and motor sizing as seen in **Figure 14** or there is some serge in current during EV driving simulation due to poor EV conversion system design as seen **Figure 15**.

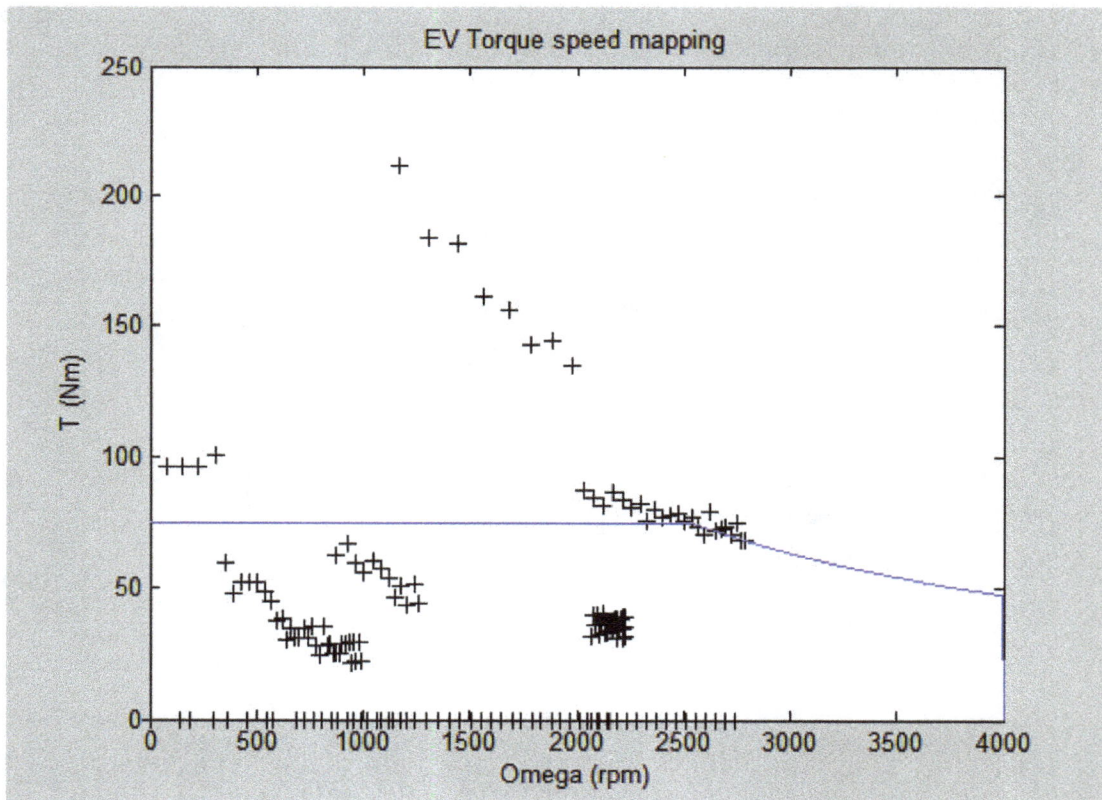

Figure 14. EV torque map results show mismatch in driving load vs. motor power.

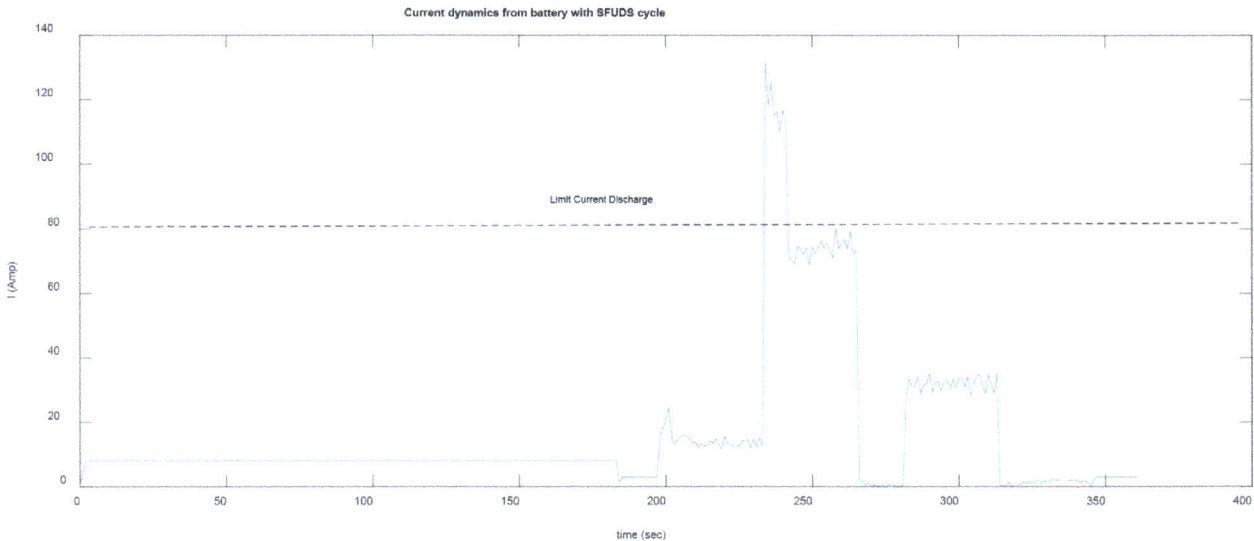

Figure 15. EV simulation show excess current drawn from batteries during some period of EV driving.

4.2. EV conversion driving range and energy storage technology

For the simplicity of modeling, PEM fuel cell with the lower heating value (LHV) plant will be reasonably modeled as a system with approximated value of efficiency [6]. Mass of liquid hydrogen supplied from cryogenic tank is calibrated to provide the same amount of energy supplied when compared to battery source. Because the fuel cell system was employed to entirely replace the batteries, regenerative braking option was not available. Comprehensive review for other types of fuel cell system and hydrogen storage for electric vehicle can be found in Larminie and Lowry's literature [4, 6]. To model the power flow model for fuel cell electric pickup truck, the battery unit is simply replaced by fuel cell system for one way flow of power since there is no regenerative braking option available. The analysis can be seen in Ref. [8].

4.3. Fault-tolerant simulation

Another important aspect for EV reliability is the fault-tolerant function. This is the safety feature when the main components of the EV, such as ECU or sensors, are malfunctioned. It ensures that EV can still provide safe operation, although in an inefficient manner. One scenario is limp home mode where EV can still be driven under limited speed and performance. By employing model-based design, limp home mode scenarios can be simulated and analyzed. As a result, accelerator redundant system can be designed and tested to compensate when fault is detected within the EV system.

4.4. Noise and vibration simulation

Improper installation or configuration of EV components could seriously induce noise and vibration, resulting in damage or shorter lifespan of EV components or undesirable noise level. EV mechanical system model can be simulated at early stage to validate the level of noise, and vibration is acceptable before the actual bench testing. Vibration characteristics such as resonance and force inducted due to rotating unbalance of the motor can be investigated and analyzed by means of model-based system design approach as well. Thus, proper noise and vibration handling method for EV system can be applied to protect and maintain EV components.

5. Conclusion

Model-based system analysis can be helpful to improve EV conversion system design, software and ECU development, testing, tuning, and diagnostic processes along with the actual EV prototype fabrication. It can also be used as a decision-making tool for future EV customization process. This methodology would benefit for EV conversion not only in terms of saving cost but also to shorten development period. As a result, reliable and low-cost EV conversion can be established.

Acknowledgements

This work cannot be accomplished without the support from several organizations. I would like to thank my engineering faculty at Naresuan University for establishing DRIVE center and providing budgets for equipment. Furthermore, I am grateful and appreciative for continuously supporting this R&D program from the corporations of EGAT (Electricity Generating Authority of Thailand) and NSTDA (National Science and Technology Development Agency).

Author details

Ananchai Ukaew

Address all correspondence to: ananchaiu@nu.ac.th

Development and Research of Innovative Vehicle Engineering (DRIVE) Center, Faculty of Engineering, Naresuan University, Thailand

References

[1] Fowler KR. What Every Engineer Should Know about Developing Real-Time Embedded Product. Boca Raton FL: CRC Press; 2008. Ch. 1-3

[2] Ukaew A. Model based system design of conceptual drive-by-wire ECU functions for electric vehicle conversion. SAE International Journal of Passenger Cars—Electronic and Electrical Systems. 2013;6(2):411-418. DOI: 10.4271/2013-01-0426

[3] Mueller S, Langjahr P. Efficient testing and cost awareness: Low-cost versus HIL-system. SAE Technical Paper 2012-01-0933. 2012. DOI: 10.4271/2012-01-0933

[4] Vuli P, Badalament M, Jaikamal V. Maximizing test asset re-use across MiL, SiL, and HiL development platforms. SAE Technical Paper 2010–01-0660. 2010. DOI: 10.4271/2010-01-0660

[5] Himmler A, Lamberg K, Beine M. Hardware-in-the-loop testing in the context of ISO 26262. SAE Technical Paper 2012-01-0035. 2012. DOI: 10.4271/2012-01-0035

[6] Larminie J, Lowry J. Electric Vehicle Technology Explained. UK: John Wiley & Sons; 2013. ISBN: 0-470-85163-5

[7] Patil K, Muli M, Zhu Z. Model-based development and production implementation of motor drive controller for hybrid electric vehicle. SAE Technical Paper 2013-01-0158. 2013. DOI: 10.4271/2013-01-0158

[8] Ukaew A. Model based system design of urban fleet electric vehicle conversion. SAE Technical Paper 2014-01-2005. 2014. DOI: 10.4271/2014-01-2005.946

[9] Ukaew A, Chauypen C. Implementation of conceptual real-time embedded functional design via drive-by-wire ECU development. International Journal of Mechanical, Aerospace, Industrial, Mechatronic and Manufacturing Engineering. 2015;9(6):940-946

[10] Choi Y, Kim Y, Moon H, Son Y. Model based design and real-time simulation of the electric bike using RT-LAB and Simulink. SAE Technical Paper 2013-01-0110. 2012. DOI: 10.4271/2013-01-0110

[11] King J, Nelson D. Model-based design of a plug-in hybrid electric vehicle control strategy. SAE Technical Paper 2013–01-1753. 2013. DOI: 10.4271/2013-01-1753

[12] Vandi G, Moro D, Ponti F, Parenti R. Vehicle dynamics modeling for real-time simulation. SAE Technical Paper 2013–24-0144. 2013. DOI: 10.4271/2013-24-0144

[13] Yeaton M. Managing the challenges of automotive embedded software development using model-based methods for design and specification. SAE Technical Paper 2004–01-0720. 2004. DOI: 10.4271/2004-01-0720

[14] William R. Real-Time Systems Development. Burlington, MA: Butterworth-Heinemann; 2006. Ch. 1–3

[15] Vincentelli AS, Zeng H, Natale MD, Marwedel P. Embedded System Development from Functional Models to Implementations. NY: Springer-Science; 2013. pp. 8-16

[16] Laplante PA. Real-Time System Design and Analysis. 3rd ed. NJ: IEEE Press, Wiley-Interscience; 2004. Ch 1, 5

[17] Lee I, Leung JYT, Son SH. Handbook of Real-time and Embedded Systems. Boca Raton, FL: Chapman & Hall/CRC; 2008. pp. 2.1-2.16

[18] Michaels L, Pagerit S, Rousseau A, Sharer P. Model-based systems engineering and control system development via virtual hardware-in-the-loop simulation. SAE Technical Paper 2010-01-2325. 2010. DOI: 10.4271/2010-01-2325

6

The Spatio-Temporal Analysis of the Use and Usability Problems of EV Workplace Charging Facilities

Eiman Elbanhawy

Abstract

With the worldwide calls to meet greenhouse gas targets and policy objectives by 2030, finding an electric vehicle (EV) on the way to work every day has become less surprising. Adapting to owning an EV is challenging to all potential users. Current users tend to rely on domestic charging for a more certain and less hassle charging opportunity. The demand is shifting towards workplace charging (WPC) as a cheap and convenient solution due to the relatively long time the car is parked there. WPC fills a critical gap in EV charging infrastructure needs by extending electric miles and building range confidence. This chapter reports on the social practice of using one of the WPC facilities in the UK. It investigates the use and usability problems that are faced (n = 12) by EV users at workplace environment in one of the UK public sector employer.

Keywords: workplace charging, electric vehicles, charging pattern, behaviour change, shared resources

1. Introduction

Transport represents one of the fastest growing sectors of the economy in terms of energy use and environmental impact. The car has become ubiquitous in late modern society; it is the spine of communities, and the leading object of mobility. Electric vehicles (EVs) show the potential to reduce the energy cost of driving [1] and the environmental burdens of the transport sector and it is seen as the core of future mobility pattern [2]. They are emerging as a zero carbon alternative to conventional internal combustion engine (ICE) vehicles. The transport sector currently supports a wide range of ubiquitous technologies [3]. Research on intelligent transport systems (ITS) covers a wide field, as it comprises combinations of communication, computer, and control technology developed and applied in transport to improve efficiency and system performance and facilitate mobility. Innovative technologies can be applied to vehicles as well as transport infrastructure and are used by stakeholders embracing transport organisation, information technology (real time information, tracking and vehicle-to-vehicle communication) and passengers to improve service quality and transport management.

A multitude of new worldwide policy objectives and measures is caused by the increasing demand for energy and electricity in particular, decrease fossil energy source stocks and the necessity to act against climate change [4]. In recent years, the environmental burden of urban road traffic has been of concern to governments and authorities of developed countries [8, 9] with an increasing interest in mitigating this [10, 11] as well as to develop and (re-) design cities to make them greener [9]. Electric mobility (E-mobility) offers considerable potential to make progress in a variety of wider environmental, societal, and economic objectives [10], which accelerates the development of smarter and sustainable cities [5, 6, 24]. Even though EVs have existed for some decades, the term is still thought of as a new technology [14]. The EV market is dynamic and has unanticipated market changes, the overall market share has grown steadily since 2011 [15], yet it is less than 1% as of 2014 [16] and in Germany and most of the other countries below 1% January 2015. The Electric Vehicle Industry Association (EVIA) forecasts varied wildly ever since and could have been affected by incentives and fuel costs, the projection was raised to 2.4% of total new vehicle sales by 2022 [15, 16]. The reasons behind the slow growth of the EV market are multifaceted. Many factors are responsible, which vary between socio-technical, economical, and psycho-temporal. In order to increase the market penetration and reach 2030 and 2050 EU and UK targets and policy objectives environmental targets [22, 14], the stakeholders particularly R&D, planning authorities and policy makers should develop deployment and operation plans. These plans should take into account the user experience and user interface, where the power of end-user feedback, design requirements and social influence are considered.

Individuals and families struggle with the decision of owning an EV, this is tied directly to the different issues related to limited range [20] and the known anxiety of not having access to a charging point (CP) [18, 19]. Research by Grahm-Rohe et al. 2012 showed that EV Drivers in England Drivers' fear that their EV will not have sufficient power to reach their next destination represents one of the main barriers to the acceptance to this emerging technology [23, 14, 16]. In England, the EV drivers reported that their experience recharging their cars is perceived simpler than anticipated [25]. Despite major technological developments in various EV areas of research, there is a list of issues that needs to be addressed. Among these, the need for a reliable and diverse charging infrastructure, which meets different user needs, is placed at the forefront [26]. Perceptions of EV-resources and, in particular, the limited range and infrastructure shifts over time differ between individuals [27].

1.1. Research importance

The chapter focuses on the emerging social behaviour of EV workplace charging facility (WPC). From literature, the WPC is gaining more attention within the EV community. Employers started to install CPs at their premises in a way to show commitment and environmental concerns, and reduce travel-to-work journeys' carbon emissions. Users started to change their patterns to include WPC within their daily schedules. However, the problem of shared resources remains especially in small-scale institutions where supply and demand problem arises (the number of the CPs is not sufficient for the current EV employees).

To be able to fully understand the nature of the EV charging at workplace environment, an in-depth study of a particular WPC scheme is presented and the emerging social practice around the sharing of CPs among staff is discussed. We derive empirical EV charging pattern and communication between users with access to the same charging network and show how they are affected by charging management as well as charging power. We started this research in early 2015, and have been reporting on some insights of the WPC practice ever since [28]. EV drivers face four particular challenges compared to drivers using the established fossil fuel infrastructure: (i) current battery technology limits range, so depending on the commuted distance drivers may be required to recharge while at work in order to get home without stopping; (ii) re-charging takes significantly longer time than conventional fossil fuel refuelling; (iii) the infrastructure for charging is poorly developed compared to the traditional infrastructure, and (iv) EV drivers may have to compete for the charge points that are available. In order to overcome these hurdles, the planning of non-domestic charging infrastructure needs to be addressed with the associated different factors.

1.2. The emerging questions

The uncertainty of having access to away from home charging facility has slowed the growing trend of smart ecosystems [18, 31, 39]. Recently, it has been claimed that WPC adds flexibility to work day, expand their comfort zone, and shall double the daily range of individuals [29, 28]. However, with the emerging need to have access to the WPC, in some cases the supply surpluses the demand, which creates sort of a competition spirit between the users, so each of them meets their mobility demand. In order to analyse the charging patterns and the importance of the WPC, research has to be backed up with evidence and empirical data that can analytically collaborate the preposition. With the lack of actual data about charging patterns [32] and user preferences, the demand of best practices and showcases is becoming an urgent matter for future development and strategic planning. This study investigates WPC systems by taking a case study approach to a functioning WPC scheme operational at a university. For this study, we had access to all possible data needed to analyse the social practice of the shared charging resources. Unlike the use of public charging, the workplace has unique characteristics:

1. Population: closed population with identified EVs usually stationary for the entire work day;

2. Equality: each EV driver has access to the same number of CPs everyone has;

3. Reachability: users can reach each other as they are working in the same place;

4. Environment: Peer pressure, human aspect, self-regulation, community collaboration, social and anti-social qualities may arise.

This chapter illuminates two research questions: Can we categorise the charging patterns of WPC users? and What are the WPC design requirements that would overcome the use and usability problems?

1.3. EV recharging alternatives

At home charging, which is referred to as the "main charging" [33], allows most EV drivers to fully charge from empty within 3–7 hours (e.g. current Nissan LEAF battery 24–30 kWh capacity, which will charge in about 3 hours at 32A or 7 hours at 16A). The time taken for an EV battery to be charged depends (charging time) on the initial state of charge (at arrival), the battery capacity, age of the battery, type, and level of CP.

Charging at home using a standard 13A outlet or dedicated 16–32A unit has positive values: (i) quiet operation, (ii) zero tail-pipe emissions, (iii) possible green energy in case of solar panel, (iv) ease of use, and (v) the car typically spends most of its idle time at this location (especially at night). However, the current limited range of many EVs means that even a full overnight charge may not meet the daily mileage demand [34]. The recharging time takes several hours, which is considered as a limitation that severely hinders the usability of EVs [35]. Moreover, domestic charging is not an option in the case of 'garage orphans'. A 'garage orphan' is defined as a resident, who has no off-street parking at their residence, or who has off-street parking, but no access to electricity supply close enough to the parking location [36].

1.4. Non-domestic charging

Besides charging at home, EV users can also make use of public CPs with more of these appearing as a facility offered by larger stores, in city centres or located at petrol stations. In the UK, CPs are usually IEC 62196 (Mennekes) sockets with a maximum of 32A (6.6 kW), although some standard 3 pin 13A CPs remain, so they are compatible with all EVs. There are four main EV charging types: slow (up to 3.3 kW), which is best suited for 6–8 hours overnight; Fast (7–22 kW), which can fully recharge some models in 3–4 hours; and Rapid AC and DC (43–50 kW), which are able to provide an 80% charge in around 30 minutes. Service providers are expanding their rapid charging networks in order to alleviate this gap in the system and to meet the end user charging demand and preference especially with the current EV market penetration. There are national charging network websites (www.Zipmap,.ac.uk), Twitter hash tag (#UKCharge) and blogs (unofficial Nissan Leaf forum (www.leaftalk.co.uk) that show updates for all CPs across the UK. In addition, each service provider [e.g. Charge Your Car (CYC) in the North East of England] and Chargemaster Plc. covering mainly Midlands and South of England have their own websites and live maps to show the charging network updates, see **Figure 1**. In addition to other countrywide open source applications, see **Figure 2**.

1.5. Workplace charging

WPC has often been considered in third place in terms of priority behind domestic and public charging infrastructure, and has not been given broad attention. WPC refers to EV charging facility that is provided at or near the user's place of employment [37]. Users only decide to charge away from home for specific reasons due to extra required planning and extra costs (Jabeen et al., 2013). On the other hand, the availability of WPC eases the pressure of electricity peak demand on the grid [38, 31]. Although not all users need to charge at work in order to return home, the ability to charge increases flexibility [33] and fills a critical gap in the EV

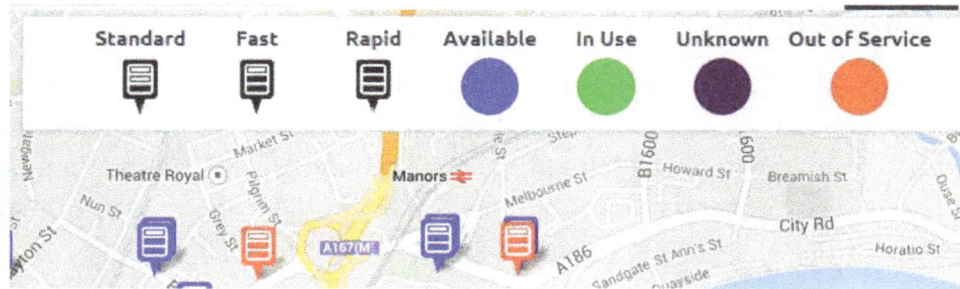

Figure 1. Print screen of CYC network real-time update.

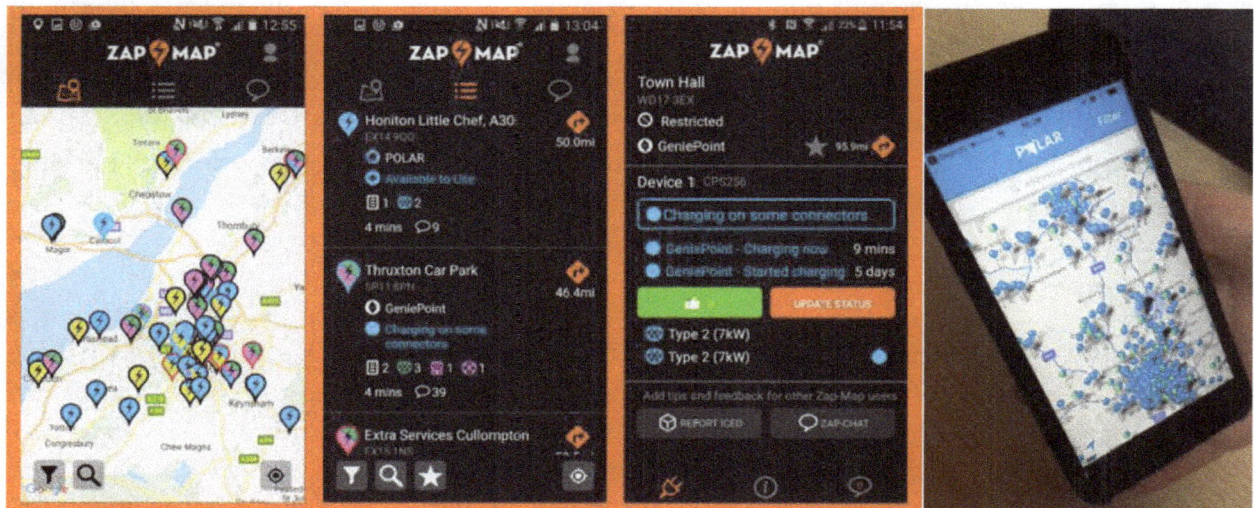

Figure 2. ZAP map and polar, smart phone applications for EV charging network real-time updates.

charging infrastructure [19]. In return, EV might be more spread out and peak problems could be reduced [39–41]. It increases the certainty of having access to charge apart from the domestic one, which in turn decreases EV range anxiety. Because EVs are parked at workplaces for substantial periods over the weekdays, WPC is a promising option only if practical ways can be found to provide the needed infrastructure.

1.6. E-mobility from an end user perspective

Inside many EVs, there is a built in telematics service that allows the driver to interact with the car's energy management systems, as in Nissan Leaf telematics system, NissanConnect EV (Carwings before) mobile application and car display (**Figure 3**). In addition to the user interface, the user can access the car remotely from their smartphone or computer by downloading the application or logging in with their credentials. To check the public charging network, there are various smartphone applications that can be downloaded.

EV owners often create their own collection of applications that cover their mobility demand (e.g., users who do not charge non-domestically, are not keen to install various charging-related applications). Based on the charging network memberships and the open access maps

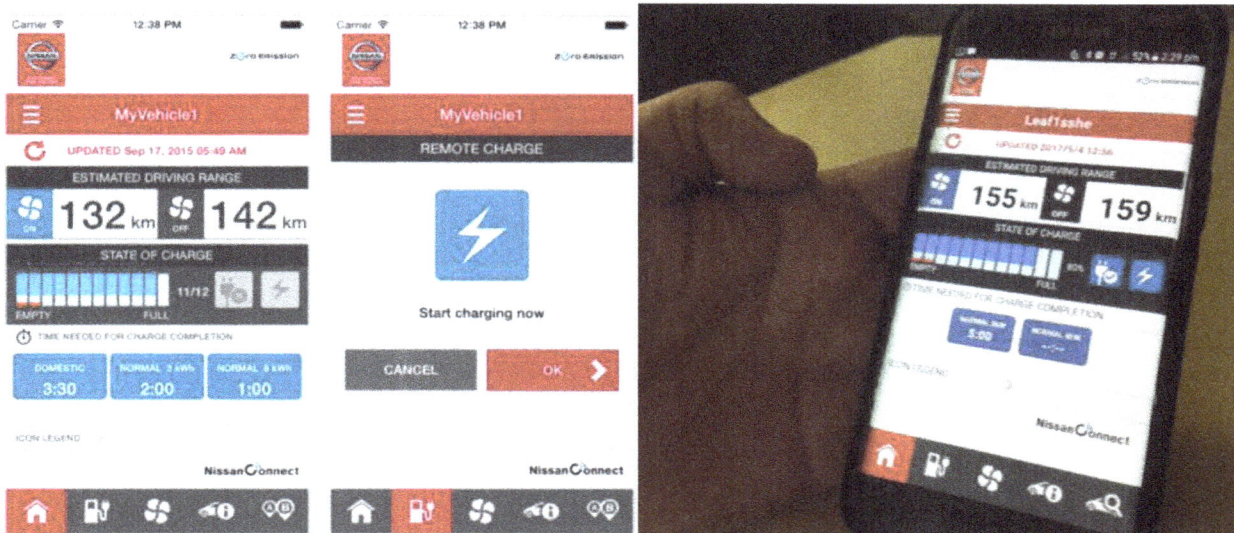

Figure 3. Nissan leaf telematics–NISSANCONNECT EV user interface-UI.

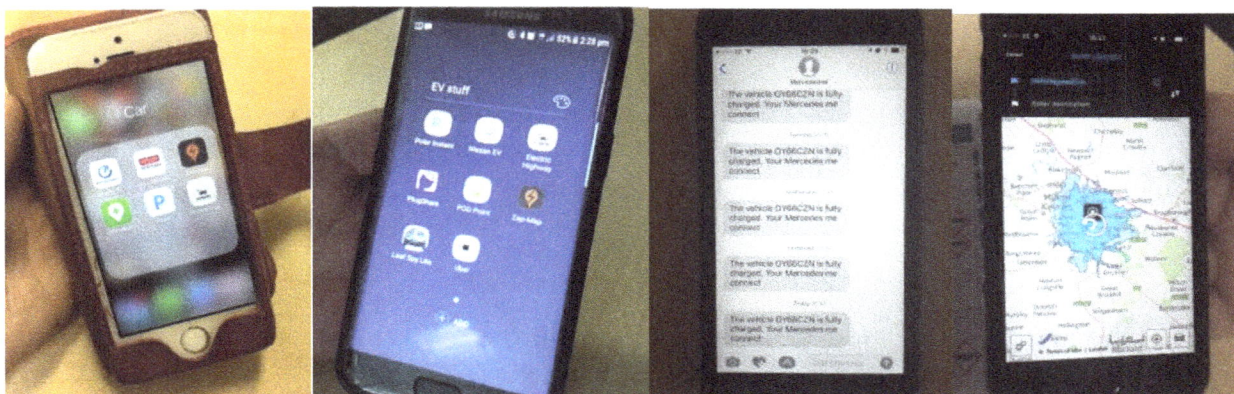

Figure 4. Various collections of smartphone applications EV users have.

each EV driver is using, they each create a list of applications; **Figure 4** gives examples for different sets of applications 4 EV users have on their smart phones.

There are national charging network websites (www.Zipmap,.ac.uk), Twitter hash tag (#UKCharge) and blogs (unofficial Nissan Leaf forum (www.leaftalk.co.uk)) that show updates for all CPs across the UK. The social media and the service providers' applications do not cover the privately owned recharging network, which includes the CPs allocated for particular community, company staff members or customers only. This takes us to the definition of available CPs the EV driver may find in their journeys. An available CP means:

1. It is working and does not report any fault;

2. It is free, not occupied by other drivers or no queuing;

3. It has a compatible power socket;

4. The driver can access it (e.g. finding a CYC charging point and the driver is not registered with CYC);

5. It is physically located and the driver is aware of its existence. The awareness and the certainty of having a CP is the main issue of WPC, which is the third alternative.

WPC refers to an EV charging facility that is provided at or near the user's place of employment [31]. Because EVs are parked at workplaces for substantial periods of time during weekdays, WPC is a promising option. Although not all users need to charge at work in order to return home, it is another chance to have access to away from home charging, which in turn increases flexibility [4, 19] and decreases EV range anxiety.

2. Literature and related work

Since 2012, studies and research projects have tackled EV adoption [10], range issues, infrastructure [15], charging loads on electricity grid [19, 41] and the use of renewable energy to charge the battery [40]. WPC practices are published as business cases or corporate reports [42] by service providers or a summary of a governmental initiative or grant or scheme announcements [20, 19, 26, 28]. These reports provide data regarding the number of CPs, locations, power capacity, service provider, billing policy, and the number of users. They lack information about operation and management, as there are no examples or practices of usability and utilisation of resources.

2.1. In the UK

In 2013, the Secretary of State for Transport announced a series of grant schemes for plug-in vehicle charge points via the Office for Low Emission Vehicles (OLEV). The grant schemes would provide up to 75% towards the cost of installing new CPs for public sector bodies to install WPC on their estate. In total, 43 public bodies were successful and were granted various amounts (£3200.00 up to £237,000.00) in 2013 and 2014 to install CPs [28]. In the case of councils, the grant covers public CPs across the city. For example in The Open University, two CPs were installed for use by eight EV fleets. In the city of Milton Keynes, 10 staff members leased a Nissan Leaf for 18 months and were given access to three dual WPC for free. The council reported 1500 charging events for the first year with a positive EV user feedback accessing WPC. A further five CPs were installed on three sites of a major health care trust. In another part of the country, 49 fast chargers were installed and in a major city in the south, 20 dual CPs across the city were installed with different service providers in order to deal with the memberships and billing policies.

2.2. In the USA

The emissions from commuters are the main driver giving workplace charging facilities a higher priority. According to [35] in Las Anglos, average employees commute over 24 miles to work generating three times the emissions of the county fleet operation (428,000 MTCO2).

For that, the county requires employers with 250 employees at a site to provide charging facilities for alternative means of transport. Some studies reported some statistics on anticipated use of the network based on surveys. In one California survey, 37% of EV drivers had access to WPC [31]. California is an active region in installing WPC. In 2011, 20 case studies were released [29]. The case studies are all in California and they vary between small-scale companies. Seven companies are intermediate and four large-scale firms reaching 500 EV users. Same as in the UK, these case studies were presented in various business reports showing the installation cost and the service provider with no details on the operation and the charging practice.

In 2013, a workplace challenge was launched by the U.S. Department of Energy covering 40 states. The partners were committed to assess their employee's demand of EV charging at the workplace and to develop and execute a plan to provide charging access. In total, 150 employers joined the challenge and some of them were providing green energy to top up the cars (charging point powered by solar energy).

From 20 case studies that were reported, the majority (80%) provides the charging for free. The billing policy was designed based on the number of the EV users commuting to the workplace. It is free of charge for companies with the capacity of (1–67 EV users) otherwise it is fee-based charging (per kWh or per hour). The common themes of these studies are the billing policy and the ratio between the number of the CPs and the EV users. The companies compete with each other to get a better environmental corporate image. Firms and public bodies install CPs at their premises to promote to carbon emissions of their employees. The UX dealing with the WPC is not a priority.

2.3. WPC trials and showcase

2.3.1. Shell technology centre

In 2010, an EV trial was carried out and reported. The study was to investigate the social influence regarding consumer perceptions and preferences in a technology-based workplace, Shell Technology Centre in Thornton, England. A total of 57 staff members out of 500 (medium-sized workplace) were involved in the trial and 21 out of which were interviewed. The aim of this trial was not to investigate the WPC practice; it was to examine the anticipated social influence and consumer preference for alternative means of transport. Two EVs were offered to the participants to try driving and sharing the experience with their co-workers [2]. The workplace environment was selected, as was the exposure of a limited number of staff to an actual EV served to stimulate conversations with co-workers who did not participate in the trial. It also helped exploring the social influence, peer pressure and the effect of word-of-mouth on the end user perception and preference of driving an EV.

2.3.2. Future transport system

In 2013, Future Transport Systems Ltd. launched a trial (as part of the SwitchEV project) in Northumbria University. In total, (n = 12) staff members participated in the EV trial to drive an EV for 2 months, which aimed at depicting the charging patterns of the participants having access to two onsite CPs [28]. Each user was given an EV (Nissan Leaf) to personally use with free access to charging on campuses. In Northumbria University, the 12 staff members joined based on the voluntarily-based selection process. The users signed a consent form to allow the research team accessing their car's telematics service (Carwings). The study aimed at raising awareness and informing people about EVs and available charging facilities around the two campuses. The trial was followed by a survey capturing respondents' feedback and their EV perception before and after the trial [34].

Nevertheless the trials recruited a good number of participants and the responses were positive, the trials are not an indicator of a successful real practice. The trials can show potentials and forecasted performance or behaviour. Due to the short term of the trial and with the less exposure to driving an EV on a daily basis, understanding the WPC practice remains a non-disclosed matter. With the lack of actual data about charging patterns and user preferences, the demand of best practices and showcases is becoming an urgent matter for WPC future development and strategic planning.

2.3.3. A WPC showcase in the USA

Unlike the case studies reported in California by [6, 29], Connor, 2014 [8] talked about the lesson learned of having a WPC for 6 years in the company he works for in Oregon, USA. Connor is a solar and electric vehicle blogger; he shared the missteps of having WPC. This practice was reported informally in a blog with an EV advocate, who was keen to share their experience with WPC. In 2008, the second trial of installing and operating a WPC took place (after the first trial in late 90's). Four 3.3 kw chargers were installed and eight employees were having EVs. Charging was for free and the slots were 50% of the time utilised. In 2011, more users joined the EV market and the charging became more difficult. The employer ran a pilot study to analyse the system and see the best strategic management approach to operate the WPC. The existing CPs were removed and new six 6.6 kW chargers and the employer applied a charging fee, $1 per hour. This shift badly affected the utilisation of the WPC dropping to 50% use and the CPs became available. Charging the employees for using the CPs added an element of discipline to the WPC system. Users were keen to go and unplug their EVs once they are done with charging allowing others to use the bay. By 2012, the pilot was over and the charging data was collected and analysed. The employer released the new scheme, which is: Payment is per kWh (8 cent per kWh) not by hour, charging using WPC was 20% cheaper than domestic charging and four CPs were added to have in total 11 WPC serving 60 EVs.

With the new scheme, there was no incentive to move the car once done with charging, and the users were back to bay blocking. By time, and since it is a workplace and as it is a closed population with known working hours, the EV users started to know each other. They decided to have a dashboard cards so they can contact each other and they started a company-wide forum to discuss issues. This WPC is an example of a self-regulated AM/PM shifts for charging events. The community started to grow, with some occasions of spotting an EV plugged in but not charging (using the bay to park only). With the peer pressure, these occasions became rare. With some technical problems, a software glitch was responsible to set 5 of the 11 CPs free, and the users were back to bay blocking.

2.4. Gap in the literature

The WPC context provides several unique opportunities for innovative research on social influence and preferences formation. There is a clear gap in the literature due to the lack of information about the WPC practices, access to charging data, and meeting WPC users. Assessing real world use and monitoring of a WPC system will allow the communities (user and provider) design requirements to emerge. Ethnographic studies and spatio-temporal data analytics that explore and identify the behaviour of EV users rather than their perceptions or attitude will allow actual relevant insights to emerge. Relying on anticipated behaviour or probabilistic scenarios of WPC environment would be misleading due to the special nature of this system compared to the public shared charging network and to conventional refuelling infrastructure. EV users do not have pre-existing preferences for novel attributes that they have not previously experienced [1]. Thus, some behaviours are spontaneous and constructed in the process of facing new technology [7] and allow users to examine technology affordances [38].

The latter case study illustrated the mechanism of the WPC as it was seen by one of the EV users. An insider EV user critiqued the missteps and drawbacks of the system. With a bottom up approach, an informal scheme was formed regulating the WPC system. One of the issues the EV users experienced was a practice called bay blocking. It is the phenomenon of someone, who has finished charging but has not moved their vehicle so that another user can use the CP. This phenomenon rarely occurs in publicly available CPs as the service provider charges the EV driver as long as the car is occupying the bay (as long as it is plugged in). This phenomenon influentially affects the use of WPC. The present study will be analysed and compared to this showcase interrogating any similarities between the two systems.

3. Methodology

The present study applies an observational research method; it is a longitudinal study in which a case study was identified and the data was gathered over a period of time [43]. This study takes a data-driven approach. Following a multi method approach, we elicited details about EV users' perceptions, social practice, interactions and charging patterns sharing the WPC at a UK research institute. The methodology was phased, spanning 12 months (March 2015–March 2016), see **Figure 5**. Due to the nature of the longitudinal study, it was expected that new employees join the WPC scheme while others opt out. The availability of all users throughout the phased study was not guaranteed.

Table 1 summarises the methods used and the involved participants in each stage. The first stage of the study was a structured interview [44] (Interview I), which was carried out in March 2015. In total, 4:07 hours were spent interviewing the (n = 10) participants. Qualitative and quantitative data was collected through open-ended questions. **Table 2** summarises the EV users' profiles and relevant data. In 2015, there were 12 users registered in the WPC scheme; however, only 10 were regularly using it. In 2016, three users joined the system and one opted out to have in total 13 EV owners.

Methodology

[1]	Interview (Round 1)	*March 2015*
[2]	EV Diary (Round 1)	*March 2015*
[3]	Tracking Charging Events	*January2015-*
[4]	Tracking Charging Notifications	*March2016* *[14 months]*
[5]	EV Workplace Diary	*February-March 2016 [One month]*
[6]	Data Visualisation of [3] and [4] *Spatiotemporal Analysis*	*January2015- March2016 [14 months]*
[7]	Comparison Analysis Between Diary 1 and Diary 2	*March 2016*
[8]	Interview (Round 2) of [5] and [6] *Participatory Data Analysis*	*March 2016 [One Week]*

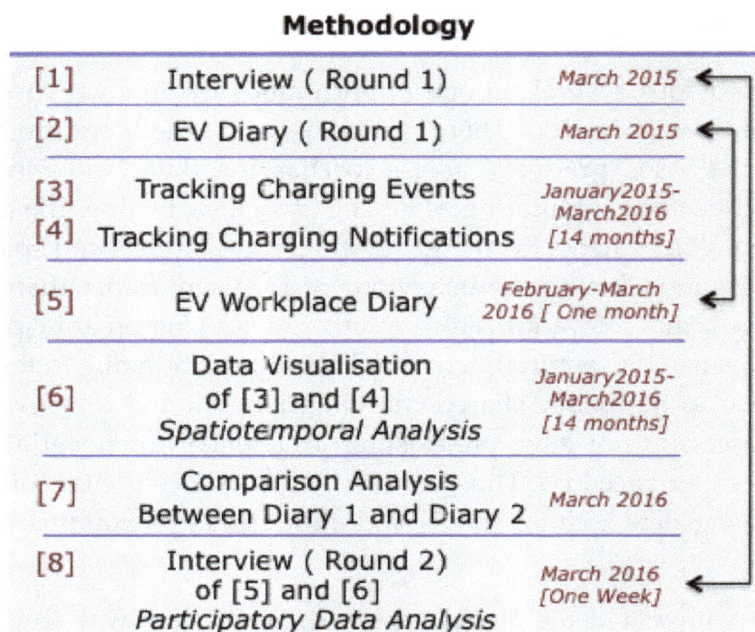

Figure 5. Methodology diagram showing the study timeframe.

Method	Participants
[1] Interview 1	EV01, EV02, EV03, EV04, EV05, EV06, EV07, EV08, EV09, EV10
[2] EV diary 1	EV01, EV02, EV03, EV05, EV07, EV08, EV09
[3, 4]	All users
[5] EV diary 2	EV01, EV02, EV05, EV07, EV08, EV09, EV10
[6] Data visualisation	EV01, EV05, EV07, EV08, EV09, EV10, EV11,EV12,EV13
[8] Interview 2	EV01, EV05, EV07, EV08, EV09, EV10, EV11,EV12,EV13

Table 1. Study phased methods and the involved participants.

In order to get factual figures about the state of the battery and the use of the WPC, participants were also asked to fill in a diary with details of date of the charging event, SOC on arrival, start and end times, and the SOC after charging. The EV diary took place in March 2015 for three consecutive weeks, where (n = 7) filled it in and returned it. In order to interrogate the charging practice, data analytics were conducted accessing the database with charging events provided by the service provider. The database included the (CP identification code, data of the charging event, starting time within the day, end time within the day and the overall kW consumed). In addition to monitoring the notifications, the EV users send charging updates to each other.

In February–March 2016, another EV dairy was designed to report individuals WPC charging pattern. The responses of the two dairies were compared to investigate the changes in user charging patterns during 1 year of use. To have clear insights and justification of users' behaviour, a data visualisation technique was deployed showing each charging event and the

User	EV model	Gender	Purchase date	Distance to work
EV1	Nissan leaf	Male	54 month	3 mile
EV2	Nissan leaf 2nd hand	Male	26 month	6 mile
EV3	LEAF	Female	55 month	8 mile
EV4	LEAF	Male	30 month	4 mile
EV5	LEAF	Male	19 month	5 mile
EV6	LEAF	Male	15 month	6 mile
EV7	LEAF	Male	18 month	30 mile
EV8	LEAF	Female	16 month	27 mile
EV9	Zoe	Female	18 month	8.5 mile
EV10	Zoe	Female	13 month	3 mile
EV11	Zoe	Female	3 month	3 mile
EV12	Nissan leaf	Male	6 month	5 mile
EV13	Hybrid	Male	4 month	60 mile

Table 2. Participants personal and EV-related information (The first user – EV01) joined in August 2014 and the newest joined in December 2014 (EV10). Some of the participants are sharing their car with their spouses (EV01, EV04 and EV11).

corresponding notifications sent by the user. The visualisation was used over the second interview (March 2016) as a Participatory Data Analysis (PDA) approach [45]. PDA aims to understand the human behaviour by conducting interviews in which participants interpret, justify and reflect on their own data. Devereux was the first to highlight that reflection on such personal ways of reacting can be used as a source of knowledge [46]. Interview II was semi-structured and recorded, 7:08 hours in total were spent interviewing (n = 9). As a qualitative research method, a thematic analysis was carried out of each interview to identify and report patterns (themes) within the data collected and helps interpreting various aspects of the research questions [47].

3.1. Case study

The present case study covers a UK university as a one of the WPC grant successful bids. In August 2014, two CPs were installed, see **Figure 6**. Each CP has two ports and each port has a capacity of 32 A, in two different locations on campus. One CP is in the visitor car park near main reception to accommodate visitors and staff (site 1) while the other is in a staff car park (site 2). Since installation, a utility management company was contracted to monitor the WPC. The billing policy is that visitors, as per request, could use the system free of charge (they are loaned an RFID card). As for the staff, providing free electricity (fuel) would be considered a taxable benefit. To avoid this, an annual membership fee was charged for their cards. Users need an RFID card, which is free for visitors but costs GBP 30 per year for staff. Once the car is plugged in, the green light flashes showing that the CP is "charging", and once the charging is over, the CP keeps flashing, "charged". If the CP is not in used, it will have blue colour. **Figure 6a–e** shows the other side of the cable while being plugged in the car, which is controlled by the EV users.

Figure 6. (a)–(e). On-site WPC facilities and charging compatibility.

The WPC is operated on a de-facto first-come, first-serve (FCFS) scheduling protocol [48]. Those who come later have to wait for the charging service in a queue [49]. However, in this case, EV users do not have a system to register and queue; they have to wait for undefined period of time when someone is done with charging so they can swap. Weeks after establishing the charging network and having more users joining the EV scheme, the users realised that the lack of communication between all the users was a barrier to efficient sharing of the charging facilities. They created a simple mean of communication to facilitate the charging process on campus. They agreed among themselves that when anyone started or stopped charging, they would send a notification message to the mailing list that indicated the estimated time for charging, see **Figure 7**.

It all started with an email with a subject line: EV – On charge. When, we, the employer added another charging bay, the users started to indicate this in their notifications, free spaces. In 2016, the waiting bay was added in each site and users started to send notification if they are waiting, see **Figure 8**.

By the time, an EV community was informally formed, and a friendly atmosphere began to appear. Not long after, more users joined the scheme, and the demand superseded the supply,

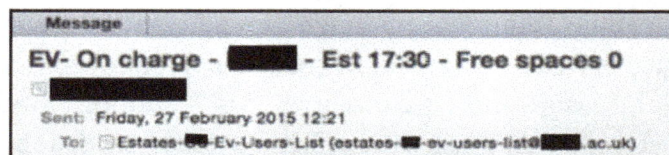

Figure 7. Printscreen of CYC network real-time update.

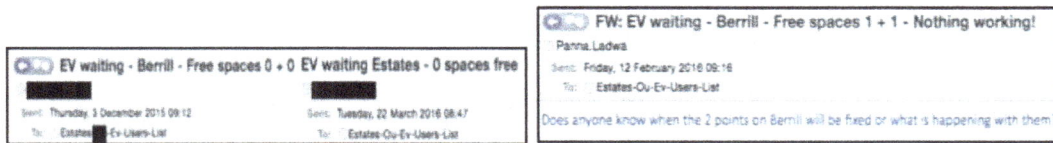

Figure 8. Printscreens of EV user notifications.

Figure 9. (a)–(d). Various notifications sent by EV users.

adding stress and competitive behaviour to the recharging process. Users started to send polite reminders asking others, who are bay blocking, to remove their EVs. Taking a community centric approach, the users created a Google-shared document, where they can log in their charging sessions. Initiatives and suggestions were brought up by the users trying to find solutions to release some of the recharging stress, **Figures 9a–d**.

3.2. 2015 interview (round 1)

Through Interview I, the users (n = 10) were asked 24 questions classified into 5 categories: (i) EV profile, (ii) driving profile, (iii) workplace insights, (iv) HCI in EV context and (v) usability. The questions varied between qualitative and quantitative addressing questions related to their motivations of purchasing an EV, charging and driving patterns for work-based trips, WPC practice, use feedback, communication system and design recommendations. Attention was given to their insights and feedback on the use of the WPC. The questions that addressed the perceptions of shared resources were:

- Does the current system allow you to meet your mobility demand?

- Is anyone taking a priority to charge? And if any, do you see this as a fair protocol of managing the WPC?

- Do you see any shortcomings in the service provided? Any recommendation for an improvement?

The social influence plays a main role in the adoption and usability of EVs [29, 28]. The motivations varied between environmental concerns, the habit of being a technology geek, long-term based financial calculations, the self-satisfaction of being an early adopters or a risk taker. The interviews showed the two main predictors of the purchase decision: domestic and WPC access. With no hesitation, the users said:

"Having access to domestic charging is compulsory; otherwise, owing an EV would not have been possible." [EV07]

"I am waiting for my domestic charger to be installed next month, I only rely on the WPC for now." [EV10]

"Some EV users rely mainly on workplaces, yet domestic charging is essential for non-work, weekends, and long journeys." [EV04]

WPC is a cheaper, more accessible (closed population) and convenient (vehicles are parked typically for at least 8 hours during the day) option compared to public charging. Through interviews, the EV users commented on the ad-hoc email list solution they had created.

"It is a good temporal mean of communication. Surely, it will break down with more users and CPs, it does not scale." [EV04]

"Driving an EV is a joy; however, the system is associated with socio-technical, behavioural needs. The communication between the CPs and us deemed fundamental." [E0V2]

"It opens a channel of communication; however, it is not a platform with real-time updates." [EV10]

"I do not have a smart phone, I come early after I drop off my kids at school, and charge almost everyday morning." [EV05]

As for the shared resources, the responses were different.

"The current practice now is based on first come first served. I am ok with that, as I have to be at work at 8 AM everyday. Wondering what others think!" [E0V9]

"Honestly, 90% of my charging events are opportunistic!. I don't have to charge to secure my home journey" [EV01]

"last week I was going to the Heathrow and I was in need to charge at the workplace, I had to send notification 2 days before asking if the Estate Department can secure a bay for me, but I didn't get a response!.. Luckily, I found a bay on the day but I was there at 7 AM!!!." [EV03]

3.3. Data analytics (stage I)

Studying the various degrees of exposure to EV driving and technology, and the social network structure created by the community, sheds light on the WPC system dynamics and the emerging social practice. The first round of the interviews provided useful and meaningful insights. Visualising users' mobility patterns helped with understanding the daily patterns, types of the journeys, denoting the daily locations of individuals (mobility demand). The EV is the second household car to all participants apart from one participant, EV10.

The interviews showed user perceptions and personal preferences; however, there was a need of actual figures that reflect how people rely on the WPC. The spatio-temporal analysis explored the charging patterns, shifts, usability and users' behaviour. Data retrieved from the CPs do not indicate the real need of charging, as it does not reflect the urgency of charging event. The data does not show if the charging event was opportunistic (the EV owner plugged in their car as the bay was available) or it was urgent to have a secured journey back home.

3.3.1. Diaries

The first diary study was carried out asking the participants (n = 7) to report their state of charge (SoC) for 3 weeks. The users reported the charging sessions' timing, which was compared to the CP database for validation. In April 2015, the dairies were collected and responses were tabulated showing minimum and maximum SOC on arrival: EV02 scored the least SOC of 9%.

The second diary study was more detailed as it was asking the users about their arrival time on campus and if they needed charging everyday they are in. The users filled the diary; some of them did this manually, while others preferred to fill it in a spread sheet. Among the two diaries, there were five users in common, EV01, EV02, EV07, EV09 and EV10. The responses were compared to interrogate if there was any difference in the SOC **Figure 10**.

3.3.2. Participatory data analysis

We conducted interviews with each participating EV user, in which we used the visualisations to help participants reflect on their previous charging patterns (2015 and 2016 diary studies). It

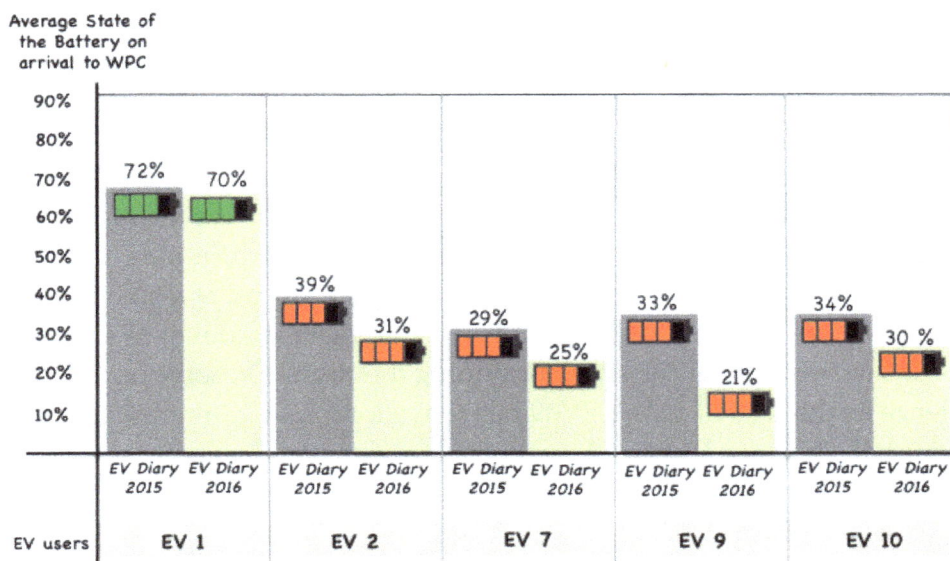

Figure 10. The average of SOC on arrival (2015/2016).

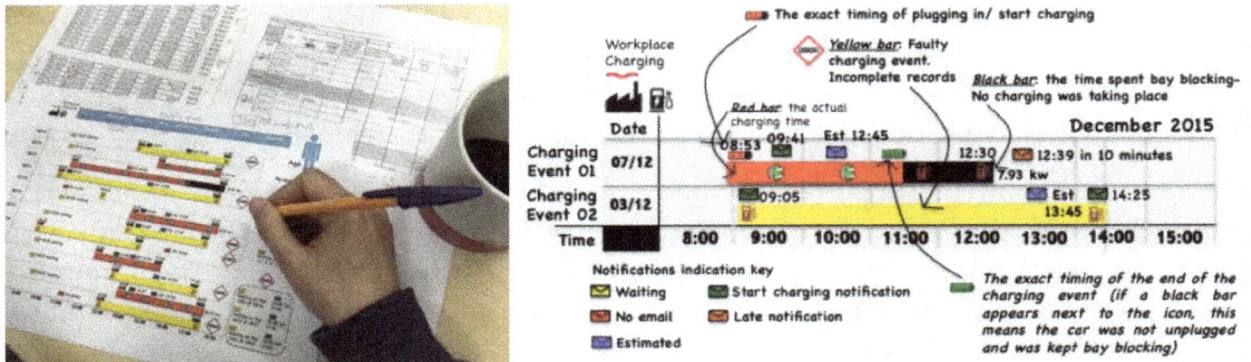

Figure 11. (a), (b). Visualising of charging even and showing it to the participants over Interview I.

was not necessary for participants to recall each day we presented, the main aim of the discussion was to observe the EV users' justification for the collective behaviour being denoted in the visualisation. Some charging events were striking and the participants were able to figure why such behaviour was made by checking their personal diaries or remembering a particular event that happened on that day.

"Only once, suddenly it didn't charge...only 50%." [EV07]

Each interview lasted about 50–60 minutes and took place at the participants' workplace. Each session was voice recorded and transcribed. We briefly introduced the visualisation to the participants through an enlarged and linear bar chart representing the charging events with all needed data, see **Figure 11a** and **b**.

The interviews were conducted using 2 A4 paper sheets with their 4th quarter of 2015 charging events and a separate sheet representing the last month (15 February–15th March) of driving and charging behaviour before the interview date.

3.4. Data participatory approach

3.4.1. Community action

The participants were asked: (1) when they send notifications to other users and (2) what kind of information they are sending over. The questions were asked over the visualisation of each user, see **Figure 7**. Some participants were very keen to send notifications to others; however, there was always a delay and in some occasions they forget to send at all. Only one participant accesses work email from their smartphones. As a result, there is a delay, as they have to return to their desks to send email after they plug or unplug their cars. On some occasions, this causes an inconvenience as the case of EV09. EV09 has to walk around 15-minute to reach site 1 and 20 minutes to reach site 2, see **Figure 12**.

3.4.2. Confidence level and use of WPC

The users were asked about the importance of WPC in their daily charging demand after another year of having access to it. They were asked whether they rely more now on WPC or

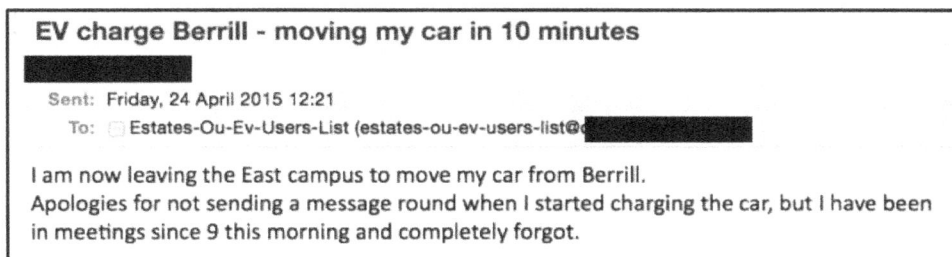

EV charge Berrill - moving my car in 10 minutes

▮▮▮▮▮▮

Sent: Friday, 24 April 2015 12:21

To: ☐ Estates-Ou-Ev-Users-List (estates-ou-ev-users-list@d▮▮▮▮▮▮▮▮▮)

I am now leaving the East campus to move my car from Berrill.
Apologies for not sending a message round when I started charging the car, but I have been in meetings since 9 this morning and completely forgot.

Figure 12. EV09 message sent to the EV users.

they plan to opt out the scheme and rely on domestic charging. The question was asked, as by looking into the charging data retrieved from the CPs, there was variance in the total kWs used by each EV user. For the charging records of 2015, EV01, EV09 and EV10 charged equivalent to <30 GBP; whereas, EV02 charged equivalent to 60 GBP and EV05, EV07 and EV08 used kW worth over 120 GBP. The participants were asked to report their car mileage since they have joined the EV scheme.

3.4.3. Charging patterns

The participants were asked about their visualised charging events and whether there is a routine, they have to stick to or they change their working hours to accommodate charging. Some users cannot make it to the workplace before 9:15 AM as they drop their kids at school. Some of those users live 30 miles away from work or the school is 20 miles away. Those users are aware that by the time they arrive, they will hardly find an available charging bay. From the charging records, those people target Noontime charging sessions. EV05 was saying:

> *"I do not look for a charging bay now when I arrive, I go straight to my office. Once I am on my desk, I login and keep an eye on the nearest estimated time. I mark the email with a 10 minute late reminder…I go to plug it in." [EV05]*

Other users give up if they could not find a morning slot as they will be too busy to go over the day to plug it in and unplug it.

> *"I usually have enough in my battery to go home. So if I miss the morning session, I don't bother charging." [EV05]*

Other users charge only twice a week, on Tuesdays and Fridays and with the FCFS protocol, it is becoming a hassle.

> *"Now realising how the charging affects my day and considering one of the charging bay wasn't working… it costs me a lot of effort to charge here."![EV09] "I don't live very far from work; however, the way we use the car is: you take the car, you do the school run." [EV10]*

While most of the users are competing to charge their EVs, some users gave up their slots for those who in need to charge.

> "At work, I used to charge it more often, but I know there's a queue and because I can charge it at home. In fact I did charge it yesterday and it is on charge at the moment. Only because I had to go out lunchtime, so I come back and there was a free space..." [EV12]

3.4.4. Bay blocking

Bay blocking (black bar in visualisations) appeared for all participants. Some participants had short bars indicating that they were stuck in a meeting or the time it took them to go to their car. In other cases, the black bars were almost the same length of the red bars (charging), which means 50% of the time the car was plugged in, and was not charging.

> "I was in a meeting so I couldn't unplug it earlier." [EV10] "We are here to work not to charge the car. I am not supposed to shift a meeting to charge the car." [EV05] "I look on the dashboard how many hours for a full charge. I walk to my office, I check my calendar, and based on it, I send the estimated time. And if there is a! meeting, I will definitely extend the time and go afterwards."[EV09]

3.4.5. System reflection

The participants were asked about their opinion about the system as some of them have been using it for more than a year and others just joined early this year. The users' attitude towards the system is different and a few of them felt unwelcomed because they have hybrid cars or live close to the university.

> "If decided to have an exact charging session, lets say...a 3 hour session from (8,30-11,30), I will use it lesser." [EV09]

> "I try to be early to charge before others." [EV07]

> "Sixty miles away from here. Without any traffic, travelling at 70 miles an hour, it takes me one hour to get here on the M25 and M1. I have to leave home with a full battery. I get here, in winter, with about 15% left. In winter I need a good five hours. If I don't get the five hours I can't get home." [EV13]

Throughout the interviews, the users found a chance to freely discuss and reflect on the current system. Some practical solutions were suggested.

> "Yes. It was my idea especially when we were down to one post that we limited to an hour and a half a week per person– people said, "No, I need to charge three hours a day." [EV12]

> "The charging here is a big stress, a major stress...I need 3 hrs to charge my battery!" [EV13]

3.5. Data analytics (stage II)

Throughout the study, certain personality traits emerged, which can identify the individuals using a WPC, some of which are social (sending on-time notifications), whereas other traits are anti-social (bay blocking). Via interpreting the records reported in the dairies and the critical analysis of the interviews' responses, a WPC user will have at least three qualities of the following seven, as one quality in each criterion: mentality, priority and requirements. A WPC user maybe a considerate person, who cares about other EV owners sharing the limited resources with them or a selfish user, who cares about his convenience and charging needs. Each WPC user would also set a priority; whether they will put their work as a priority (e.g. they will not change anything in their work habit to accommodate shared charging), they will not shift a meeting or reschedule it to avoid bay blocking; however, they will try to avoid this as much as they can and they will show remorse for doing so. The third trait is that they will not stay in the system if it adds more pressure to their daily life and will leave the WPC scheme and go back to rely mainly on domestic charging. The last criterion is the requirements. Each WPC user would have a vision in mind for the ideal fair and efficient WPC system. These requirements vary; some users demand advanced ubiquitous computing, looking for interactive mobile applications that bring ease to their charging process; whereas, others still prefer an intuitive WPC system that requires less computer involvement.

3.6. Observations and discussion(s)

When studying WPC for emerging vehicle technology, neglecting social influence processes will ignore or underestimate the potential for consumer perceptions to develop and shift. Analysing WPC systems shows cases of variant consumer personas, charging behaviour, and the need of the WPC as part of the daily routine. Charging behaviour is the collective behaviour, the EV user performs. It is the charging session they try to get, the arrival time, the time they send updates to others, the state of charge, the state of the charging point, and the time they are plugged in versus the actual charging time. It is the emergent behaviour of sharing WPC. Thus, based on this study, we cannot confidently approve the preposition of "WPC adds flexibility to work day and shall double the daily range". The WPC infrastructure is intrinsic to certain EV users, who are ready to deal with the social dynamics associated with it. A WPC system is not a prototype of a public charging network at a smaller scale, and it is a subset of an e-mobility system that has its own unique paradigms, policies and conditions.

3.6.1. A. RQ1: The four WPC user personas

In a previous study that was carried out covering different sample sizes of EV users [50], three personas were created: the old school, the risk taker and the opportunistic. The old school is the EV user, who is still afraid to have a flat battery so they are over protective and very conservative in their charging pattern. The risk taker is the EV user, who extends their driving range by exploring irregular road trips relying on public CPs. The opportunistic are the EV driver, who does not own an EV but drive fleet EV and promote car sharing as a way to be environmentally aware. These personas fit typical EV users; however, the WPC users have different personas due to various factors: shared resources protocol, peer pressure, selfishness, competition mode, priorities of individuals and work-related arrangements, which are associated with the charging process. The human aspect and the social interactions created different personas that can identity different WPC users. Four personas are generated from exploring the WPC social practice, see **Figure 13**.

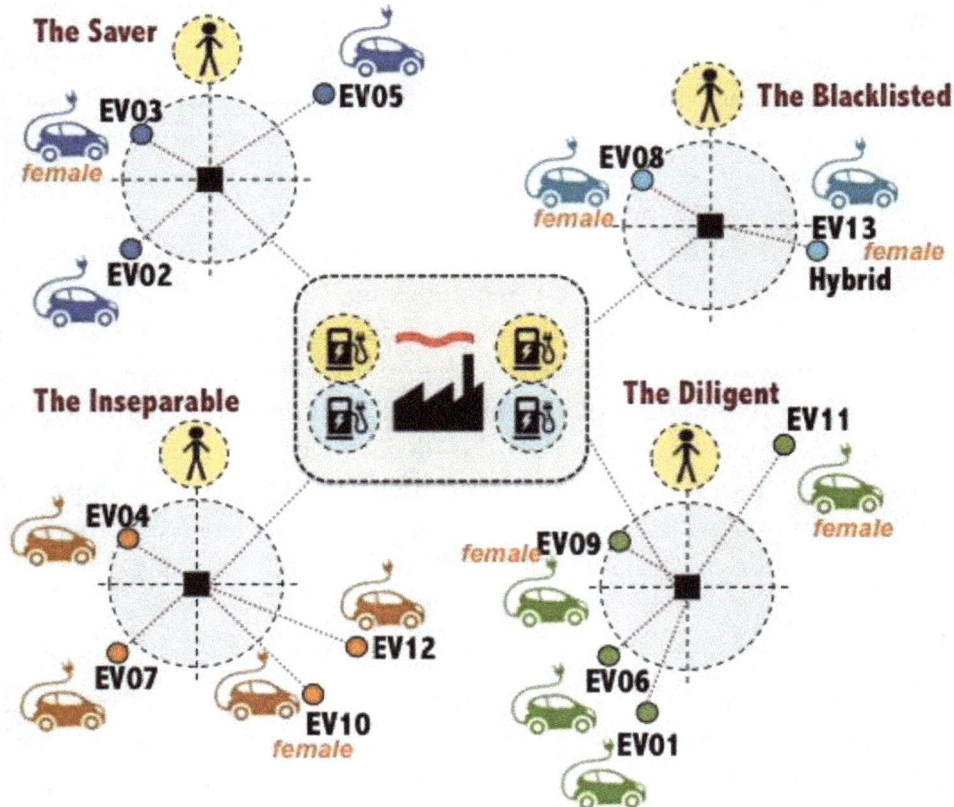

Figure 13. EV WPC personas.

3.6.1.1. The inseparable

This persona describes the EV users (EV04, EV07, EV10 and EV12), who are very considerate to other users, who are sharing the WPC. They demand daily or at least three times a week WPC. They usually arrive at work with a relative low battery (10–20%). Due to their high demand, they make sure they deserve a secured access to WPC. Even though, the system is managed with a FCFS protocol, they manage to schedule their day to make sure they charge. They abide themselves with the communication system giving a chance to whom in demand. They notify others once they are plugged in/unplugged and may bay block in late occasions. This persona is not likable within the EV community, as they believe they take advantage due to their flexible time schedule or that they live locally, so that recharging their batteries is not as stressful to them as to others (**Figure 14**).

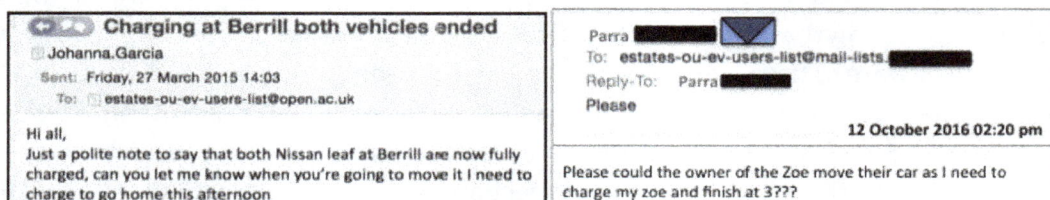

Figure 14. EV09 message sent to the EV users.

3.6.1.2. The diligent

The diligent persona is the EV user, who would try to charge at the WPC if the charging would fit within their work routine (EV01, EV06, EV09 and EV11). They are not going to change their daily routine. They arrive to the workplace with an average of (40–50%) as their charging demand is not as high as the *Decent persona*. Those users tend to be more active in the EV forum and open channels with the employer to discuss solutions and alternatives. They keep reminding other users not to bay block or to update their charging records. They are considered as system administrators.

3.6.1.3. The saver

This persona includes EV owners (EV02, EV03 and EV05), who are very close to being opportunistic users. They may not need to charge at all at the WPC. However, this persona includes risk takers, who can extend their comfort zone reaching a very low SoC. Their confidence level has increased by the time they use the car. They are confident enough to reach the workplace with one cell or even zero cells left in their batteries. Due to their local commute, they are familiar with the work journey to an extent that they can confidently leave home with only 5 miles left in their battery knowing that they will charge at the WPC. They are more flexible compared to the *Diligent persona* as they can slightly shift their work patterns to accommodate charging. They charge at their ease, they do not pay attention to the communication pool, they check if they miss morning shift.

3.6.1.4. The blacklisted

This persona includes EV owners (EV08 and EV13) whom their charging matter is not seen as urgent as others. This persona includes those who have hybrid, so relying on petrol is an option. Although they have the same right as other users, the community may stigmatised them as their case is not as critical as full electric users. Also this persona includes those who are against the suggestions for limiting the charging sessions, as this does not fit their mobility demand. They do not commute to work five times a week and live relatively far from their workplace. They do not bay block, yet they make good use of the system and charge at work very frequently. (EV08 used kW worth over 120 GBP). This persona is the least flexible compared to other personas due to mobility demand.

3.6.2. B. RQ2: Usability and related issues: recommendations for assessment and design of new WPC system

There is no definite theory that controls the use of WPC, analysing the current WPC-formulated lesson learned informing the design of future technologies that can improve the away from home charging experience:

- WPC management: sophisticated infrastructure to communicate between the users and the management is not compulsory. Simple communication pool can work. However, it has to have a way to deter bay blocking;

- Billing policy: Annual fees with unlimited charging does not serve applying self-regulated based systems;

- System assessment: Relying on the WPC retrieved data is misleading especially in the kW calculations. Interviewing EV users justifies unexplained behaviour.

- Peer pressure: may only reduce anti-social incidences.

4. Conclusion

Ev is still in an early technology adoption stage; it has not yet taken the role of main stream of major daily transportation option [51]. We cannot confidently approve the preposition of "WPC adds flexibility to work day and shall double the daily range" [9]. The WPC infrastructure is intrinsic to certain EV users, who are ready to deal with the social dynamics associated with it. A WPC system is not a prototype of a public charging network at a smaller scale, and it is a subset of an e-mobility system that has its own unique paradigms, policies and conditions. This is an ongoing research; the next step is to analyse the social practice with more users joining the scheme and testing different simulation scenarios for the scheduling protocol.

Acknowledgements

The author would like to thank MK: Smart Research Project team at the Open University for supporting this research and all EV participants who were involved throughout the study.

Author details

Eiman Elbanhawy

Address all correspondence to: elbanhawy@brookes.ac.uk

Oxford Brookes University, Oxford, UK

References

[1] Xydas E, Marmaras C, Cipcigan LM, Jenkins N, Carroll S, Barker M. A data-driven approach for characterising the charging demand of electric vehicles: A UK case study. Applied Energy. 2015;**162**:763-771

[2] Harrison G, Thiel C. An exploratory policy analysis of electric vehicle sales competition and sensitivity to infrastructure in Europe. Technological Forecasting and Social Change. 2016;**114**:165-178

[3] Beeton D. EV City Casebook_50 Big Ideas Shaping the Future of Electric Mobility. 1st ed. Vol. 2. Newcastle Upon Tyne, UK: Urban Foresight Ltd; 2014. p. 74

[4] Schäuble J, Kaschuba T, Ensslena A, Jochema P, Fichtnera W. Generating electric vehicle load profiles from empirical data of three EV fleets in Southwest Germany. Journal of Cleaner. Elsevier. 2017. DOI: 10.1016/j.jclepro.2017.02.150

[5] OLEV. Government funding for residential on-street charging for plug-in vehicles: A guide for members of the public. 2013

[6] OLEV. Making the Connection: The Plug-in Vehicle Infrastructure Strategy. United Kingdom: Office for Low Emission Vehicles (OLEV); 2011

[7] Orsato RJ. Sustainability Strategies When Does it Pay to be Green? Basingstoke: INSEAD Business Press; 2009

[8] Michaelis L. GHG mitigation in the transport sector. Energy Policy. 1996;2(10-11):969-984

[9] Breithaupt M. Towards liveable cities- international experiences. In: The Future of Mobility Options for Sustainable Transport in a Low Carbon Society. Eschborn, Germany: Expo; 2010

[10] Wee BV, Maat K, De Bont C. Improving sustainablity in urban areas. European Planning Studies. 2012;20(1):95-110

[11] Lozano AP. Intelligent energy management of electric vehicles in distribution systems short master thesis in electric power systems and high voltage engineering. Denmark; 2012

[12] Graham-Rowe E, Gardner B, Abraham C, Skippon S, Dittmar H, Hutchins R, et al. Mainstream consumers driving plug-in battery-electric and plug-in hybrid electric cars. Transportation Research. 2012;46(1):140-153

[13] Elbanhawy E, Dalton R, Thompson E. Interrogating the relation between E-mobility recharging network design and drivers' charging behaviour. In: 32nd eCAADe; 2014. p. 132

[14] Hjorthol R. Attitudes, Ownership and Use of Electric Vehicles–A Review of Literature. Institute of Transort Economics, Norwegian Centre for Transport Research. Oslo; 2013

[15] Kodjak D. Consumer Acceptance of Electric Vehicles in the US. The International Council on Clean Transportation. Washington, DC; 2012

[16] Yu Z, Li S, Tong L. Market dynamics and indirect network effects in electric vehicle diffusion. In: 52nd Annual Allerton Conference on Communication, Control and Computing; 2015. p. 11

[17] S. (South A. N. E. D. Institute. Electric Vehicle Industry Association (EVIA). 2017. [Online]. Available: http://www.sanedi.org.za/EVIA/EVIA.html

[18] Millikin M. Green Car Congress: Navigant Forecasts 18.6% CAGR for Plug-ins in North America to 2022; 2.4% New Vehicle Share in US. 2013. [Online]. Available: http://www.greencarcongress.com/2013/09/20130910-navigant.html

[19] EU. 2030 climate & energy framework. EU Action. 2014. [Online]. Available: https://ec.europa.eu/clima/policies/strategies/2030_en

[20] Bonges H, Lusk A. Addressing electric vehicle (EV) sales and range anxiety through parking layout, policy and regulation. Transportation Research Part A. 2016;83:63-73

[21] Krupa JS, Rizzo DM, Eppstein MJ, Brad LD, Gaalema DE, Lakkaraju K, et al. Analysis of a consumer survey on plug-in hybrid electric vehicles. Transportation Research Part A: Policy and Practice. 2014;**64**:14-31

[22] Li S, Tong L, Xing J, Zhou Y. The market for electric vehicles: Indirect network effects and policy design. Journal of the Association of Environmental and Resource Economists. 2015;**4**(1):89-133

[23] Nilsson M. ELVIRA, Electric Vehicle: The Phenomenon of Range Anxiety. Sweden: Lindholmen Science Park; 2011

[24] Franke T, Gunther M, Trantow M, Rauh N, Krems J. The range comfort zone of electric vehicle users–concept and assessment. IET Intelligent Transport Systems. July 2015;**9**

[25] Graham-Rowe E, Gardner B, Abraham C, Skippon S, Dittmar H, Hutchins R, et al. Mainstream consumers driving plug-in battery-electric and plug-in hybrid electric cars: A qualitative analysis of responses and evaluations. Transportation Research Part A. 2012;**46**:140-153

[26] Brenna M, Dolara A, Foiadelli F, Leva S. Urban scale photovoltaic charging stations for electric vehicles. IEEE Transactions on Sustainable Energy. 2014;**5**(4):1949-3029

[27] Egbue O, Long S. Barriers to widespread adoption of electric vehicles: An analysis of consumer attitudes and perceptions. Energy Policy. 2012;**48**:717-729

[28] Elbanhawy EY, Price B. Understanding the social practice of EV workplace charging. In: Adjunct Proceedings of the 2015 ACM International Joint Conference on Pervasive and Ubiquitous Computing and Proceedings of the 2015 ACM International Symposium on Wearable Computers (UbiComp/ISWC'15 Adjunct), Osaka, Japan. PURBA 2015. pp. 1133-1141

[29] Lindgern J, Lund P. Identifying Bottlenecks in charging infrastructure of plug-in hybrid electric vehicles through agent based traffic simulation. Low Carbon Technologies. 2015;**10**:1-9

[30] Huang Y, Zhou Y. An optimization framework for workplace charging strategies. Transportation Research Part C: Emerging Technologies. 2015;**52**:144-156

[31] Banez-Chicharro F, Latorre JM, Ramos A. Smart charging profiles for electric vehicles. Computational Management Science. 2013;**11**:87-110

[32] Yang Z, Slowik P, Lutsey N, Searle S. Principles for Effective Electric Vehicle Incentive Design. 2016

[33] Calstart. Best Practice for Workplace Charging. California; 2013

[34] Huang S, Infield D. Demand side management for domestic plug-in electric vehicles in power distribution system operation. In: 21st International Conference on Electricity Distribution; 2011

[35] Sukru Kuran M, Carneiro A, Lannone L, Kofman D, Mermoud G, Vasseur J. A smart parking lot management system for scheduling the recharging of electric vehicles. IEEE Smart Grid. 2015;**6**(6):1-1

[36] PRP. Demand for Electric Vehicle Charging Stations. Seattle; 2012

[37] Scott J, Kehoe C, Boone J, Chiladakis L, Paauwe G, Baroody J. Workplace charging. California Plug-in Electric Vehicle Collaborative. 2010. [Online]. Available: http://www.pevcollaborative.org/workplace-charging. [Accessed: 01-Jan-2015]

[38] Haq A, Cecati C, Strunz K, Abbasi E. Impact of electric vehicle charging on voltage unbalance in an urban distribution network. Intelligent Industrial Systems. 2015;1(1):51-60

[39] Kristoffersen T, Capion K, Meibom P. Optimal charging of electric vehicles in market environment. Applied Energy. 2011;88(5):1940-1948

[40] Spoelstra J. Charging Behaviour of Dutch EV Drivers. Utrecht, Netherlands: Utrecht University; 2014

[41] Spoelsra J, Helmus I. Public charging infrastructure use in the Netherlands: A rollout-strategy assessment. In: European Electric Vehicle Congress-EEVC; 2015

[42] Kettles D. Electric Vehicle Charging Technology Analysis and Standards. Florida Solar Energy Center, Florida; 2015

[43] Mann CJ. Observational research methods. Research design II: Cohort, cross sectional, and case-control studies. Emergency Medicine. 2003;20:54-60

[44] Merriam SB. Qualitative Research and Case Study Applications in Education. A joint publication of the Jossey-Bass education series and the Jossey-Bass higher and adult education series. San Francisco, CA; 1998

[45] Bergold J, Thomas S. Participatory research methods: A methodological approach in motion. Forum Qualitative Sozialforschung/Forum: Qualitative Social Research. 2012;13(1). ISSN 1438-5627. Available at: <http://www.qualitative-research.net/index.php/fqs/article/view/1801/3334>. Date accessed: 08 nov. 2018. DOI: http://dx.doi.org/10.17169/fqs-13.1.1801

[46] Devereux G. From Anxiety to Method: In the Behavioral Science. Hague, Paris: Mouton & Co; 1967. p. 396

[47] Boyatzis RE. Transforming qualitative information: Thematic analysis and code development. Thousand Oaks, CA: Sage; 1998

[48] Sensini P, Buttazzo F, Ancilotti G. Ghost: A tool for simulation and analysis of real-time scheduling algorithms. In: IEEE Real- Time Educational Workshop; 1997. pp. 42-49

[49] Vernon M, Manber U. Distributed round-robin and first-come first serve protocols and their application to multiprocessor bas arbitration. In: International Symposium on Computer Architecture; 1988

[50] Elbanhawy EY. Straight from the Horse's mouth: 'I am an electric vehicle user, I am a risk taker.' [EV14, M, c. 30]. In: EAI International Conference in Smart Urban Mobility Systems (SUMS); 2015

[51] Gao Y, Farley K, Ginart A. Safety and Efficiency of the Wireless Charging of Electric Vehicles. SAGE; 2015

7

Motion Dynamics Control of Electric Vehicles

Shinji Kajiwara

Abstract

In this chapter, I will explain the dynamics of electric vehicle and the support systems of drivers in detail, considering both structure and the function of the vehicle. Furthermore, the reliability is discussed. In car development and design that I have, car dynamic control system, turn ability, comfort, and safety must all be considered simultaneously. The safety and the comfort for the driver which are connected with various road surfaces and as well as the speed depend on the physical performance of the vehicle. In this chapter, we will explain the dynamics of the vehicle and the support system of the driver in detail, considering both the structure and function of the vehicle. In the design and development of car dynamic control system, turn ability, comfort, and safety must all be considered simultaneously. The safeness and comfort during a drive on various road surfaces and speed depend on the performance of these basic abilities of the vehicle.

Keywords: ABS, TRS, ESC, acceleration, understeer, oversteer, electric motor control

1. Introduction

Several of the greatest advances in automotive technology in the past 20 years are in the area of safety management. Due to the speed of microprocessors, downsizing, and the advances in software development, automobiles continue to evolve. In this chapter I will explain in detail for the dynamics of the vehicle and the driver's support system, considering both the structure and function of the vehicle. In recent years, science and technology competition in the automobile industry has been intensifying. The basic ability of a vehicle is "running," "turning," and "stopping." Currently, in-vehicle equipment is equipped with countless electric and electronic subsystems.

Some electric vehicles can control four wheels independently. It is necessary for the electric car to make a safe, reliable, and robust control system. Also, the four independent wheels of an electric car give the driver greater control and more freedom of movement. More than 100 times per second, the on-board computer can sample the input data from the steering wheel, acceleration pedal, brake, etc. and calculate how each wheel should respond. Because the wheels are independent, one or more wheels brakes and other wheels accelerate, which can enhance traction and motion control. That is, it is very difficult to drive without a controller.

Many of these subsystems include well-known standard systems such as antilock braking system (ABS), electronic stability control (ESC), traction control system (TCS), and so on. Besides that, I will also explain torque vectoring and active roll control. The EVs may have regenerative cooperative brake, which is equipped with the conventional friction hydraulic brake and the function to recover to the battery. The regenerative brake is recovered as electrical energy by rotating the motor with kinetic energy of the vehicle during braking. However, only regenerative brake does not provide sufficient braking force, and controllability is not good enough. Therefore, the controller of regenerative cooperative brake is configured to cooperate the braking force by the conventional hydraulic brake and the braking force by regeneration and thereby obtain the necessary braking force. The controller determines how much braking force is required. The maximum regenerative brake is applied within the range of the braking force. Then, the brake will compensate for the insufficient amount of brake.

First, I will explain the operating principles of each system. Next I will describe the components necessary to support the systems such as sensors, actuators, mechatronic subsystems, control elements, and so on. The stability and maneuverability of automobiles are essential for the drivers. The stability of the vehicle is the ability of the vehicle to return to a stable state during driving when faced with an irregular road surface condition. The maneuverability of a vehicle is the ability of the vehicle to quickly change direction during driving based on the steering of the driver. The stability and maneuverability of the vehicle can also be defined as the driving stability of the vehicle. Another major focus of automotive engineering is vehicle safety performance. As the safety performance of the vehicle improves, all drivers will be more confident than every day. In order to improve the reliability of the systems, two acceleration sensors may be used. The reason is that when one sensor breaks down, the failure can be easily detected by the other sensor. When estimating the vehicle slip angle, the vehicle lateral acceleration value is time-integrated and used for calculation. Therefore, if there is an error in the acceleration, the error is also time-integrated, and the error becomes very large.

As a whole, packaged together with ABS, ESC, and TSC, driving the vehicle equipped with these will greatly reduce the risk of fatal crash. Regardless of what kind of safety function is equipped on the vehicle, it is always important to pay attention and focus on driving tasks. The effects for sports utility vehicles (SUVs) are more substantial than the effects for ordinary passenger cars. This is the result of the SUVs' center of gravity being higher than that of other passenger cars. A high center of gravity is disadvantageous for the stability of the vehicle. ABS, ESC, and TSC are brake control systems that use pressure sensors, a yaw sensor, an acceleration sensor, wheel speed sensors, solenoids, a motor, and a microcontroller to electronically modulate individual wheel torque to improve vehicle stability. These systems must have high reliability, in addition to the desired performance. There are stringent government regulations on such systems. The more tests they conduct, the more they can say about the reliability of the system. Reliability is defined as the probability that a device will perform its intended function during a specified period under stated conditions. The vehicle lateral motion control of four-wheel-independent-drive electric vehicles (4WID-EVs) and direct yaw moment control (DYC) through in-vehicle networks is studied [1]. Fault-tolerant control strategies for electric vehicles with wheel hub motors are presented in order to maintain the directional stability of the vehicle in case of a component failure during high-speed maneuvers [2].

Several of the greatest advances in automotive technology in the past 20 years are in the area of safety. Due to the speed of microprocessors, downsizing, and advances in software development, automobiles continue to evolve. In this chapter we will explain in detail the dynamics of the vehicle and the driver's support system, considering both the structure and function of the vehicle. In recent years, science and technology competition in the automobile industry has been intensifying. The basic ability of a vehicle is "running," "turning," and "stopping." Currently, in-vehicle equipment is equipped with countless electric and electronic subsystems. Many of these subsystems include well-known standard systems such as antilock braking system (ABS), electronic stability control (ESC), traction control (TRC), and so on. Besides that, we will also explain torque vectoring and active roll control.

First, we will explain the operating principle of each system. Next we will describe the components necessary to support systems such as sensors, actuators, mechatronic subsystems, control elements, and so on.

The stability and maneuverability of automobiles are essential for automotive technology. The stability of the vehicle is the ability of the vehicle to return to a stable state during driving when faced with an irregular road surface condition. The maneuverability of a vehicle is the ability of the vehicle to quickly change direction during driving based on the steering of the driver. The stability and maneuverability of the vehicle can also be defined as the driving stability of the vehicle. Another major focus of automotive engineering is vehicle safety performance. As the safety performance of the vehicle improves, all drivers will be more confident than every day.

As a whole, packaged together with ABS, traction control, and electronic stability control (ESC), driving the vehicle equipped with these will greatly reduce the risk of fatal crash to 50%. Regardless of what kind of safety function is equipped on the vehicle, it is important to always pay attention and focus on driving tasks. The estimates of reduction of fatal vehicle crashes vary from 17 to 62%. The effects for sports utility vehicles (SUVs) are more substantial than the effects for ordinary passenger cars. This is the result of the SUVs' center of gravity being higher than that of other passenger cars. A high center of gravity is disadvantageous for the stability of the vehicle.

2. Kinds of dynamics control for electric vehicles

The vehicle dynamics control system provides a means to enhance positive and integrated safety. In order to maximize the contribution to overall road safety, improvement of penetration is necessary. This can be achieved by increasing the awareness or by regulating in law [3]. To fully exploit the possibilities of the dynamics control of vehicles, emphasis should be placed on driver groups such as beginners and the elderly. Adapting to such drivers will help reduce traffic disasters.

One of the first active assistance systems based on proprioceptive sensors was the antilock braking system (ABS) since 1978 (Bosch). A traction control system (TCS) later augmented the system. Thus, electronic stability control (ESC) is adapted to the market by combination with ABS and TCS. An optimal torque distribution scheme for the stability improvement of a distributed-driven electric vehicle is presented [4].

2.1. Antilock brake systems

The first mechanical antilock braking system (ABS) was developed for aerospace industry in 1930 [5, 6]. In 1945, the first set of ABS brakes was put on a Boeing B-47 to prevent spin out and tires from blowing. In the 1960's, rear-only ABS was adopted only for high-end automobiles, and with the rapid progress of microcomputers and electronics technology, it was widely spread in the 1980's. Today, the all-wheel ABS can be found on the majority of latest model vehicles. The ABS system aims at minimizing the braking distance while retaining steering ability during braking. The shortest braking distance can be reached when the wheels operate at the slip of maximum adhesion coefficient.

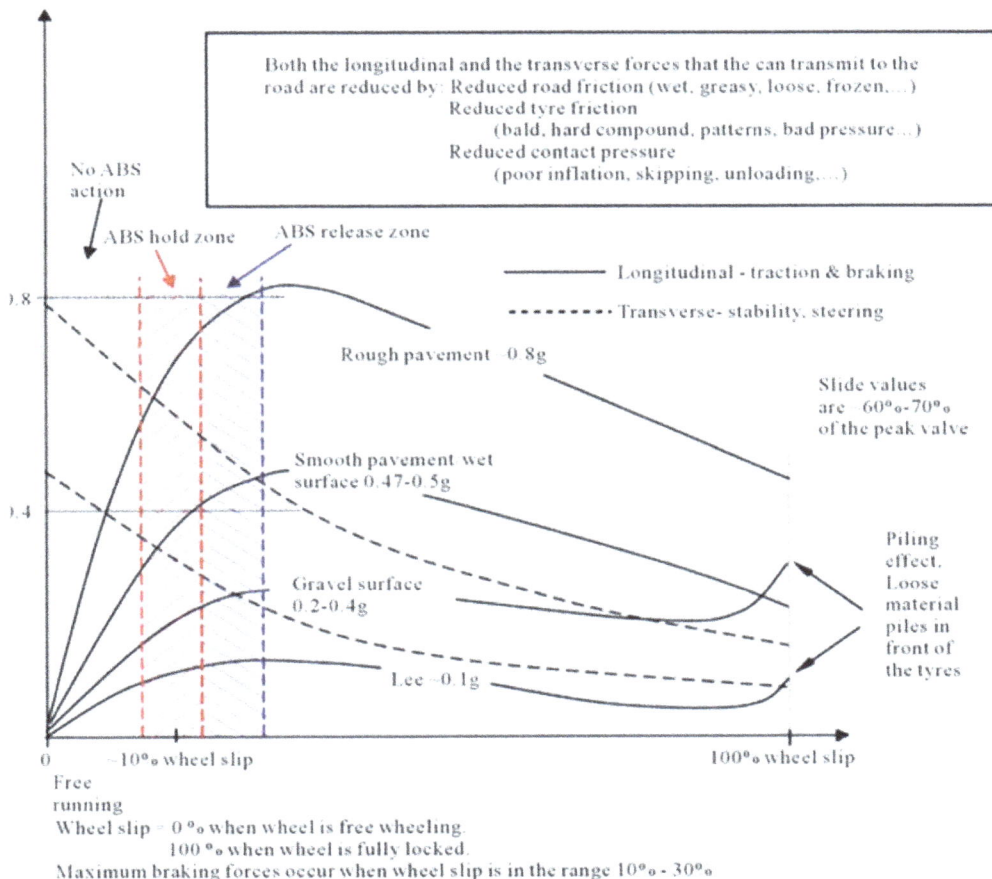

Figure 1. Relationship between braking coefficient and wheel slip [5].

The relationship between the friction coefficient and the wheel slip is shown in **Figure 1**. This figure shows that the slide value of the stop or traction force is higher in proportion to the slide value of the cornering or steering force. On applying the braking pressure, the brake torque increases. The difference between friction torque and brake torque is negative, resulting in a wheel deceleration. The wheel rotational equivalent velocity starts to decrease, and the slip starts to increase. At first, the friction coefficient increases as well as the friction torque, which narrows the torque difference. After passing the maximum friction coefficient, the friction curve changes to decrease. In order for the friction torque curve to become unstable, it might be uncontrollable in extremely high rotational wheel decelerations.

ABS is operating by monitoring the movement of each wheel using a wheel speed sensor. When ECU detects the lock of a wheel based on the speed of each wheel, by using the hydraulic pressure modulator, ECU adjusts the braking force to the desired wheel. The benefits of ABS are (1) shortening of the stopping distance, (2) improvement of stability, and (3) steering ability.

A typical antilock brake system is shown in **Figure 2**. The typical ABS components include vehicle's physical brakes, wheel speed sensors, an electronic control unit (ECU) with acceleration sensor, brake master cylinder, and a hydraulic modulator unit with pump and valves.

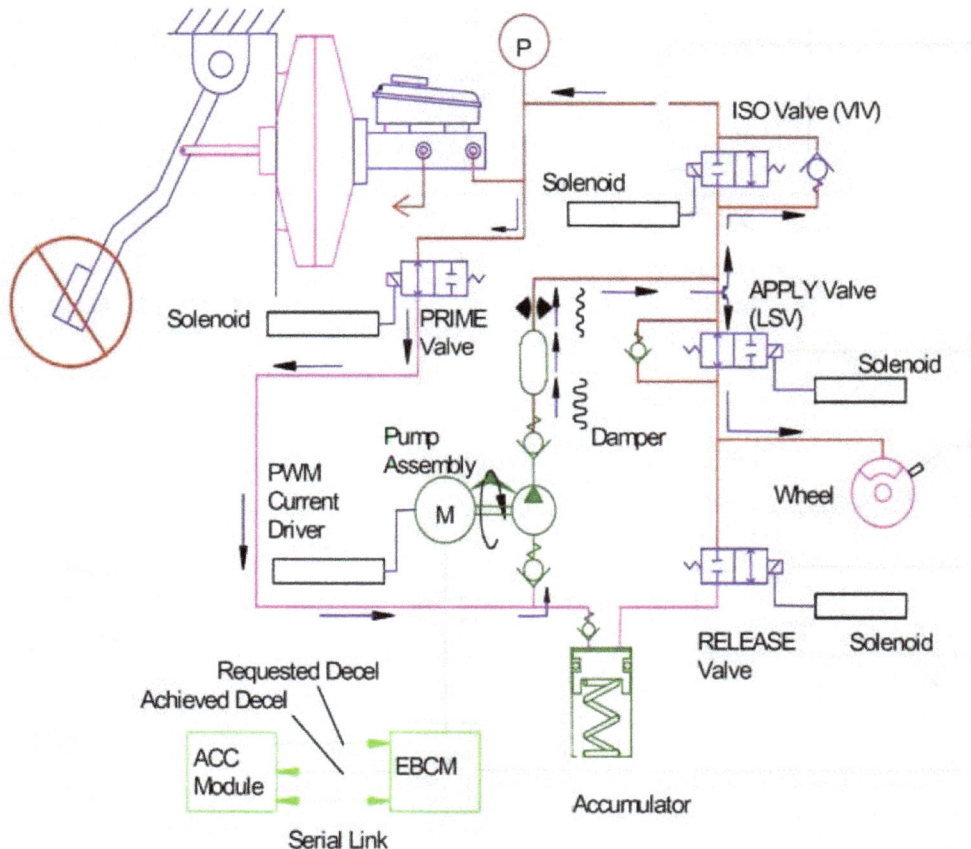

Figure 2. Antilock braking system [6].

Locked wheels provide low maneuvering force and minimum maneuvering power. As a result, the main advantage of ABS operation is to maintain directional control of the vehicle during excessive braking. In rare cases, it is possible to increase the stopping distance, but the direction control of the vehicle will be significantly bigger than if the wheels are locked up.

2.2. Control system of the ABS

ABS brake controllers pose unique challenges for designers: (a) for optimum performance, the controller must operate at an unstable equilibrium point; (b) depending on road conditions, the maximum braking torque may change over in a wide range; (c) on rough roads, the slip rate of the tire changes drastically with the bounce of the tire; (d) the change in the friction coefficient of the brake pad; and (e) the change in the braking system.

As mentioned in the previous section of this paper, the ABS consists of a conventional hydraulic brake system, in addition to an antilock component that affects the control characteristics of the ABS. ABS control has a strong nonlinearity due to the complicated relationship between the friction torque and slip rate. Another difficulty for this control problem is that the velocity of the vehicle is not directly measurable and must be estimated. The friction between road and wheel cannot be measured easily in real time or may require complex sensors.

ABS control is a very nonlinear control problem due to the complicated relationship between friction and slip. Another obstacle to this control problem is that the linear velocity of the wheel is not directly measurable and must be estimated. The friction between the road and tires cannot be easily measured, or it may require a complicated sensor.

The driver increases the braking pressure during braking. Rotational equivalent wheel speed is measured and differentiated to give wheel acceleration. The maximum friction point is passed when the wheel speed derivative falls below a predetermined threshold. In the first control cycle, even lower thresholds are applied. The braking pressure is held between a higher threshold and a lower threshold. The introduction of additional thresholds helps to suppress the effect of the final noise. When the braking pressure decreases, the wheels get speed again. When the threshold drops again, the pressure drop stops. When the wheel speed derivative exceeds the braking pressure, the pressure is increased to prevent the wheels from returning to slip values that are too small. The pressure is kept constant and rises slowly below it. When the wheel speed derivative falls again, the second control cycle begins. The brake pressure now decreases immediately without waiting for the threshold to be reached. When such a cycle is executed, the wheel rotation speed is maintained in the region where the wheel slip is close to the maximum coefficient of friction. Therefore, the braking distance can be minimized.

If the moment of inertia of the wheel is large, the wheel may lock without reaching the deceleration threshold if the coefficient of the friction is small or the brake pressure increases slowly in case of a careful braking (e.g., ice). Such a situation puts the driving ability of a vehicle at risk. Regardless of the above control cycle, therefore, the braking pressure decreases as the rotational equivalent wheel speed falls. In either condition, the maximum slip is not surpassed even when the maximum friction coefficient cannot be reached. The front wheels of the vehicle are independently controlled, and the rear wheels collectively receive lower brake pressures. This ensures operational stability. The main difficulty in designing ABS control is due to strong nonlinearity and problem uncertainty. Using the classical linear, frequency domain method, it is difficult to solve this problem and in many cases impossible. ABS system is designed around hydraulic system of system, sensor, and control electronics. These systems are dependent on each other, and different system components are compatible with minor changes in the controller software. Estimation of the coefficient of friction is necessary when maximum friction is used. It is necessary to evaluate the maximum brake in various kinds of road surface condition such as dry, ice, and snow cover. In the case of vehicles equipped with ABS, the panic braking state is ABS controlled. That is, when an ABS control cycle occurs, maximum friction is used. On the other hand, the operation of the ABS must generally be evaluated to evaluate whether the failure of the ABS system has affected the investigated accident. The prediction method employs the wheel speed measurements, in order to detect an ABS-cycle. Detecting the ABS-activity, the past two measurements of the wheel speed are utilized to extrapolate linearly a value of the current wheel speed.

2.3. Traction control system (TCS)

This section presents a scheme to enhance vehicle lateral stability with a traction control system. Traction control systems are used to prevent wheel slippage and to maximize traction forces. The control is an active vehicle safety feature designed to help vehicles make effective use of all the traction available on the road when accelerating on low-friction road surfaces. These conditions include cases where the road is wet, ice-caking, non-uniform, loose, or poorly maintained. When a vehicle without traction control tries to accelerate on a slippery road like ice, snow, gravel, the wheel becomes slippery. As a result of the wheel slipping, the tire rotates the surface of the road without grip, and the vehicle cannot accelerate. Traction control works when it detects that a wheel may slip and helps the driver to make the most of the traction available on the road. Traction control cannot produce force exceeding the maximum frictional force. In a state with little friction such as on ice, the performance with the traction control vehicle is hardly changed with the vehicle without traction control.

2.4. Control system of the TCS

The traction control system works at the opposite end of the scale from the ABS—dealing with acceleration rather than deceleration. Traction control works similarly to ABS and is often considered as a supplement to the existing ABS setups. Since many of the same principles apply to both systems, it might be best to visualize it as sort of ABS in reverse. In fact, traction control uses the same components as ABS: a wheel speed sensor that monitors the rotation speed of the front wheel or all four of the wheels, a hydraulic modulator that pumps the brake, information from the wheel speed sensor, and an electronic control unit (ECU) instructing the hydraulic modulator to pump the brake. ABS senses the slippage of the wheels at the time of braking and continuously adjusts the brake pressure to ensure maximum contact between the tire and the road surface. Some basic ABS systems require some modifications. First, older accelerator cables are usually replaced by electronic drive-by-wire connections. That is, the mechanical connections between the accelerator pedal and the throttle is unnecessary. Instead, the sensor converts the position to the accelerator pedal to an electrical signal, and a control unit similar to that used in the ABS is used. Standard ABS hydraulic modulators are also expanded to include traction control components. All these parts work together to operate the traction control system. Therefore, modern ABS and traction control systems are physically one unit, while functions are different because the ECU and the hydraulic modulator are mounted together. The ECU continuously checks whether some wheels are turning faster than other wheels. It is an indicator that the wheel is losing traction. Traction control monitors the wheels of the vehicle for potential "wheel slip." When the wheel of the vehicle slides down, the tire seems to be "rotating," but the grip on the road cannot be grasped. There is little chance of controlled acceleration when the wheels are spinning. When the traction control system senses that one or more of wheels are about to slip, it corrects the problem by applying the appropriate amount of the brake to that wheel. When a possible wheel slip is detected, the ECU instructs the hydraulic modulator to immediately brake the wheel in question quickly and reduce the rotational speed. Depending on the traction control system, there are things that reduce motor output to slipping wheels. When the wheels recover the traction, the system returns to monitoring the wheel speed and comparing the wheel speed of rotation. In vehicles with reduced motor power to control the rotation of slipping wheels, the driver may experience gas pedal pulsation while the traction control is in operation. This pulsation is normal and does not indicate that there is a problem with the traction control system.

The traction control system is often considered as a supplement existing in ABS settings. Both systems work to solve the opposite problem associated with wheel slip or wheel lock. The same components as ABS, such as wheel speed sensors (sensors measuring wheel rotation speed), hydraulic modulators (brakes if necessary), ECUs, etc., are used. To add traction control to the ABS, the control unit that calculate the best traction of each wheel and other valves to the hydraulic brake modulator are necessary. Thus, it is a relatively easy way to install traction controls on vehicles already equipped with ABS. The traction control system uses the individual wheel speed sensors to measure the difference in rotational speed of each wheel. These sensors are located on each wheel. The ECU sends a system to the hydraulic brake modulator, which is attached to the ECU upon detecting that one wheel is rotating faster than the other wheel indicating that the wheels are losing traction. Traction control systems have various ways to reduce the rotating speed of each wheel. The other systems combine wheel brakes with decelerated motor power, but some "pump" the brakes to the wheels in question. In vehicles with reduced motor power to control the rotation of slipping wheels, the driver may experience gas pedal pulsation while the traction control is in operation. This feeling is similar to the experience of pulsating the brake pedal when ABS is running. When the wheels recover the traction, the traction control system monitors the wheel speed and returns to comparing the wheel speed of rotation. The test shows that traction control is effective for reducing wheel slip when accelerating with low-friction condition, but this effect is more effective than four-wheel drive. It is remarkable in the car.

However, there is a risk that the driver performs traction control to respond to the safety system. First, the driver may not be sure of the sounds and sensations associated with the proper functioning of the TCS. When it is operating, the traction control system generates a crushing sound, and the gas pedal may pulsate. If the driver is not familiar with the traction control mechanism, there is a possibility that the driver may misunderstand that there is a possibility that the traction control may be canceled due to some defects in the traction control. Secondly, the traction control helps to maintain the steering force by preventing rotation of the wheels, so the driver must be careful not to exaggerate the steering command under the condition that the traction control is working. Traction control works just by preventing the tires from slipping, allowing the wheels to fully utilize the traction available on the road. Traction control cannot increase the total amount of available traction. In order to explain this point, if the road is completely freezing and there is no traction on the road, nobody has the traction power. Traction control can be useless in case of no friction. The traction control hinders the rotation of the wheel, so the driver with traction control may accelerate normally under low traction situations. Furthermore, since the wheels of the vehicle are not rotating, the driver can maintain the steering control. However, it cannot be exaggerated that a driver with traction control does not experience more traction. If slight traction is available, better handling is possible. Therefore, the driver is advised to limit or avoid driving under slippery low-friction conditions with or without traction control.

The traction control system can allow the vehicle to reach a higher speed in low-grip conditions. It is necessary to monitor the vehicle speed and to be careful not to exceed safety speed limit that depends on the situation. Regardless of the safety system installed in the vehicle, it is important to concentrate on driving always without hesitation. By selectively applying the brakes to the slipping wheel, TCS will be able to increase the traction of the wheels. This is especially important when the wheels are on surfaces with varying levels of friction. In addition to using a brake to control the tractive force during acceleration, TCS also controls the vehicle by using the motor management. TCS system, by communicating with the motor power controller, may control the amount of torque transmitted to the wheels. If the wheel is almost no traction with the road and was simply detected by the system, TCS system will be able to significantly reduce the torque supplied to the wheels. The motor power management system adjusts significantly the amount of torque supplied to each wheels. Due to this control system, the stability of the vehicle is greatly improved.

2.5. Electronic stability control

One of the biggest technological advances that enabled in a system, such as the ESC, is a wide availability of automotive electronics. Such an electronic steering angle sensor, a wheel sensor, a lateral acceleration sensor, a yaw rate sensor, and a power controller made it possible to develop a vehicle stability system. In addition, the ability to connect these electronic components to a high-speed network helped to develop these vehicle systems.

Electronic components stabilize the vehicle in a crisis situation, preventing the control from steering. System operates through a series of high-speed digital messages sent to the control unit from the sensors which is sent finally to the actuator. At a few milliseconds, these components begin their remedial action by evaluating the condition of the vehicle, determining the necessary remedial actions, braking, and controlling the torque [7]. The basic control algorithm is to evaluate the behavior of the vehicle, determine whether it is necessary to make any changes to the vehicle, and control the components responsible for changing the behavior of the vehicle, the entire electronic component using the data. Basic ESC control algorithm for four-wheel drive vehicle is shown in **Figure 3**. The main purpose of the control algorithm is to compare the actual behavior of the vehicle and desired behavior and to determine what needs to be done in order to equalize the behavior.

The electronic system has greatly improved the safety and the performance of the vehicle. In such a system, there are some inherent limitations and problems, such as the increased risk that system designers must deal with. Because these systems are important for safety, special procedures must be performed to ensure system reliability and fault tolerance.

Finally, there are many advantages of the system, such as ESC. With the introduction of many electronic controls there are significant improvements and benefits to the vehicle system, but with this development risks and problems also arise. When the vehicle electronic systems such as ESC have been developed, there is a need to solve the problems related to the reduction of the driver's control and consumer awareness.

2.6. Control system of the ESC

ESC consists of ECU and sensors such as wheel speed sensor, steering angle sensor, yaw rate sensor, lateral acceleration sensor, motor power controller, transmission controller, accelerator pedal sensor, and brake pressure sensor. The ECU collects data from the sensors and uses that data to calculate the necessary adjustments for the brake pressure or motor torque. ESC requires many different sensor data to determine the actual state of the vehicle and the desired state of the vehicle based on driver input. To determine the speed of the vehicle and the necessary torque to each wheel, the system uses a wheel that requires speed sensors. Information from wheel speed sensors is used to calculate the actual vehicle speed and the amount of slippage each wheel is experiencing. The amount of longitudinal tire slip is calculated by comparing the wheel speed and the actual vehicle speed. The actual vehicle speed can be determined by the evaluation of the speed of non-driven wheels. In the case of all-wheel-driven vehicles,

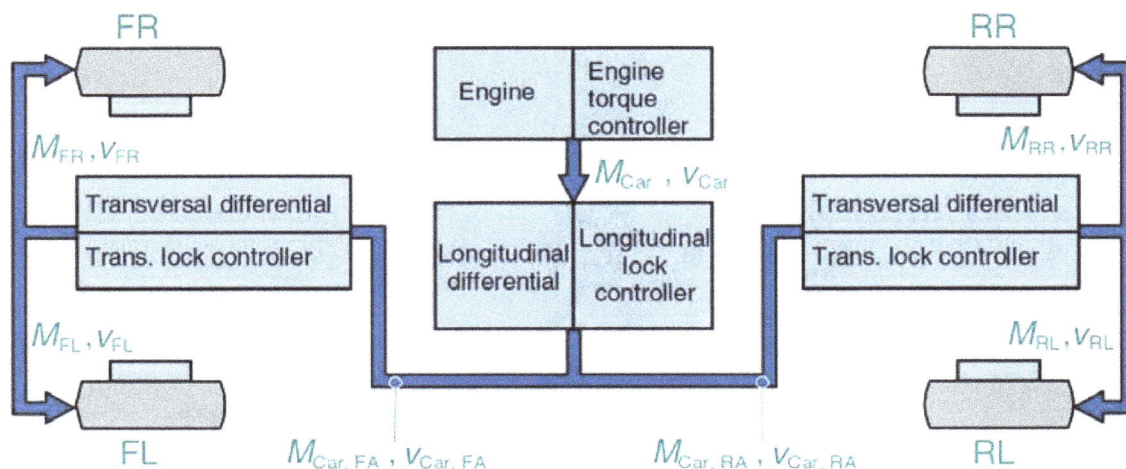

Figure 3. Basic ESC control algorithm for four-wheel drive vehicle [7].

the actual vehicle speed cannot be measured. Therefore, among the four-wheel speed, the fastest one is selected, and a pseudo-vehicle speed signal is created based on this. It is called "the principle of the select high."

The system also needs to be able to monitor the desired movement of the vehicle from the driver's input. A steering angle sensor and an accelerator pedal sensor are used to measure the input of the driver. The steering angle sensor is often mounted behind the steering wheel on the top of the steering column and measures the position of the steering wheel. The gas pedal sensor is used for the desired speed acceleration of the vehicle. Also ESC system requires the measurement of vehicle rotational acceleration around the vertical axis, which is called yaw rate. The rotational velocity is the yaw rate of the vehicle. By using the data from the yaw rate sensor, ECU determines whether the vehicle is overspeed, that is, in the oversteered state or understeered state when the vehicle is not rotating enough. The yaw rate sensor is based on a vibrating cylinder gyrometer. The metal cylinder vibrates at a constant frequency. The node is displaced according to the amount of rotation of the vehicle due to the Coriolis effect. In addition to the aforementioned sensors, vehicle stability system also uses the lateral acceleration sensor and the brake pressure sensor. ECU evaluates where the driver is steering the vehicle and where the vehicle is actually.

When the vehicle becomes unstable due to oversteer or understeer, the ESC system individually adjusts the power to the wheel. The ESC can stabilize the vehicle and control the vehicle in dangerous situations according to the slippage and torque of each wheel. When the system detects an oversteer condition while turning clockwise, braking is applied to the left front wheel. When the system detects an understeer condition, braking is applied to the right rear wheel. A schematic diagram of the control is shown in **Figure 4**. This action will help the driver avoid serious accidents caused by slippage and shaking. In addition to

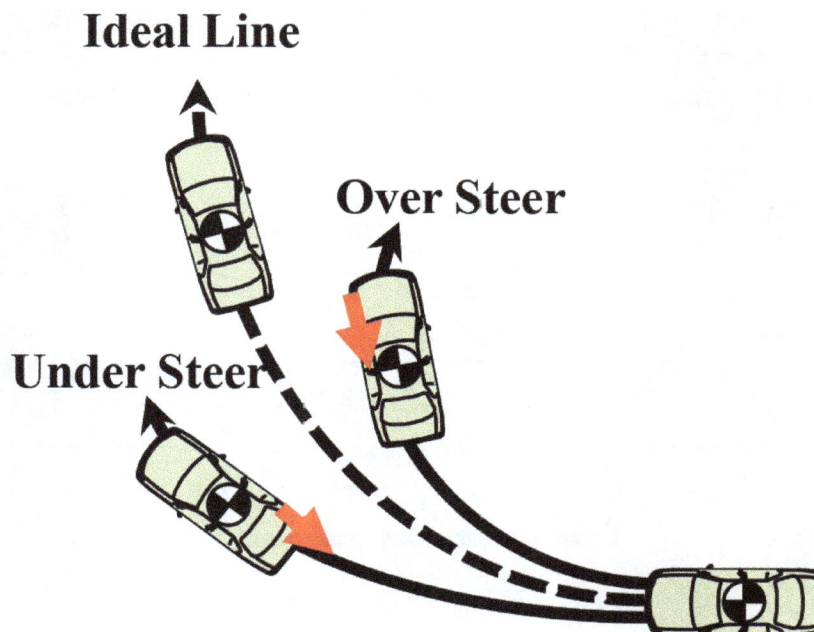

Figure 4. Schematic view of ESC method.

controlling the vehicle by applying brake, the system also interacts with the power management system by communicating with the power control unit. The motor power controller is also an embedded system that handles tasks such as fuel injection, spark timing, throttle control, air-fuel ratio, and so on. By sending a signal to the motor power control unit, the system controls the torque of the wheels. Even when the driver steps on the accelerator, the system may decide that it is necessary to reduce the torque of the wheels in order to stabilize the vehicle.

The system samples the input signal, and it is necessary to determine the current state of the vehicle. When the system detects a serious condition, the driver starts responding before the driver recognizes that the vehicle is out of control. Control time is the most important, as delays in the response of the system can have disastrous consequences for the driver. When traveling on an expressway in a frozen or damp condition, a time of one tenth of a second means the difference between striking the guardrail and safely staying in the lane. Thus, ESC must meet stringent timing requirements during operation. The main value of ESC is the ability to react and stabilize to the vehicle faster than humans. This is far superior to the fastest human response of 300 ms. Failure to meet the response time requirement of the system can result in disastrous results for both the driver of the vehicle and the other drivers on the road. In some cases, if the deadline of timing is not kept, the occupant may be killed or seriously injured. Therefore, ESC is hard real-time system that requires special precautions on systems that are designed to ensure the ability of the system to meet response time requirements.

The flow of control of ESC begins from the inputs and the current state of the vehicle driver. Steering angle, yaw rate, wheel speed, lateral acceleration, accelerator pedal pressure, the brake pedal pressure, the brake fluid pressure, and motor power management information are collected from sensors and sent to ECUs. From this information, ECUs will calculate the actual operation of the vehicle. In addition to calculating the actual movement of the vehicle, ECU must also determine the nominal or desired movement of the vehicle. The nominal motion mainly input information of the driver of the vehicle, i.e., a steering angle sensor, a throttle position sensor, is calculated based on the brake pedal sensor. The ESP design aims primarily at inducing the nominal movement of the vehicle, i.e., dynamic behavior obtained when neither front nor rear tires are saturated in the condition like a slippery road surface. The ESC controller compares the nominal value of the vehicle motion with the actual value of the vehicle motion. Once the difference between these two values is determined, the ESC controller determines the correction of the tire slip value necessary to achieve the nominal value of the vehicle motion. The overall control of the vehicle is based on the slip values of all four wheels. Tire slip is the difference between the actual speed and the vehicle tire. If the tire is rotating faster than the moving vehicle, the tire slips. The controller determines control signals that need to be sent to actuators (valves, pumps, etc.) of the system to achieve the desired tire slip value. The slip controller sends these control signals to the various actuators to change the movement of the tires.

There are several important safety issues that play an important role in the design and evaluation in hard real-time systems. One of the most important problems is the integration and verification of sensors, ECUs, and mechanical and hydraulic components. In many cases, components are manufactured by various suppliers and may have internal software that must adapt to interface with their own interfaces or other components. It is necessary to confirm the operation of integrating them. In addition, some electronic modules are general-purpose products, not designed exclusively for applications.

Their operation must be verified, and their interface must be adapted to this system. For this reason, in order to meet the stringent safety and timing requirements of the system, you must have a test and verification of each component. Another challenge facing the system suppliers is variations of vehicles. Since the driving characteristics and other components can vary greatly depending on the vehicle, the system must be reconfigured and verified for each vehicle. In addition, because system failures can cause such disastrous consequences, measures are taken to ensure partial system function and even if a failure occurs. Instead of completely off the system in the event of failure, certain important elements, such as ABS, continue to function. It is called a recovery condition. For example, if the system finds a failure in the power management interface, the portion of the system would be interrupted, and then the system will operate only during the brakes. Such steps have been done to enhance the robustness and fault tolerance of the system. One of the important functions found in ESC software is model-based sensor monitoring. This type of sensor monitoring can detect failures that are slightly out of specification. Model-based sensor monitoring operates by comparing the actual output of the sensor with the predicted output calculated by the software model. The software tracks the history of sensor output and determines the possibility of the current reading based on the model. If the controller detects a sensor operating outside the specification, the controller can reduce its own level of sensitivity to that sensor. In other words, when the sensor is under suspicion of failure, the magnitude of the system's response to that sensor can decrease dramatically.

ESC improves the ability of a car to steer on a frozen or slippery road significantly, also prevents vehicles from uncontrollable spin during critical situations, and provides many benefits to vehicle performance and safety. Furthermore, it is possible to minimize the adverse effects when there is a possibility on slippery roads that are presumed recognized beforehand—but when the drivers cannot safely control the vehicle, stabilization will be difficult. In case driver is losing control of the car, ESC can react quickly with brake and motor power intervention at speeds more than ten times faster than humans. ESC, when viewed as a mechanism for the driver's support in a crisis situation, is highly profitable and greatly improves vehicle safety. There are several important safety issues that play an important role in the design and evaluation in hard real-time systems. One of the most important problems is the integration and verification of sensors, ECUs, hydraulic components, etc. In many cases, the components are manufactured by various suppliers and may have internal software that must adapt to interface with their own interfaces or other components. The carmakers need to integrate them and check the operation. These were not intended to replace the driver's functionality. The control of the vehicle is ultimately the responsibility of the driver before the autonomous driving advent. In order to avoid potentially fatal incidents, it is important for consumers to be notified of system that functions and has restrictions like ESC.

2.7. Torque vectoring control (TVC)

Directing the yaw moment control by acceleration and by the brake is an effective technique to positively prevent accidents, which is widely spread today. The world's first straight-ahead yaw moment control system on the right and left torque vectoring type was developed and announced in 1996. This system directly controls the yaw moment when acting on the vehicle regardless of whether the vehicle is accelerating or decelerating by transmitting torque between the left and right wheels. It is an operation essentially blending braking, steering, and accelerating to provide drivers and passengers with a more accurate and stable feel. In order to uniformly control these forces and to optimize the vertical load of each tire to realize a smooth and efficient vehicle, the motor torque and the braking force are adjusted according to the steering input. Increase the vertical load of the front tire by causing a slight deceleration to achieve the natural cornering attitude of the car. Basically, it monitors three parameters of vehicle speed, throttle position, and steering wheel speed. In a system that uses the difference between the left and right torques, the torque distribution of all of the four wheels are controlled [8]. A schematic of this control is shown in **Figure 5**.

This system showed that the movement of the steering wheel was reduced, the first turning moment of each maneuver was more accurate, and fewer modifications were made in the middle. The steering performance under almost all conditions at almost all speeds has been improved. Also, on a slippery surface, the difference becomes more prominent. TVS closely monitors the speed of the steering wheel input, and it gives a signal to the motor to slightly lower the torque. This function improves cornering performance at all stages, from normal operations to critical operations.

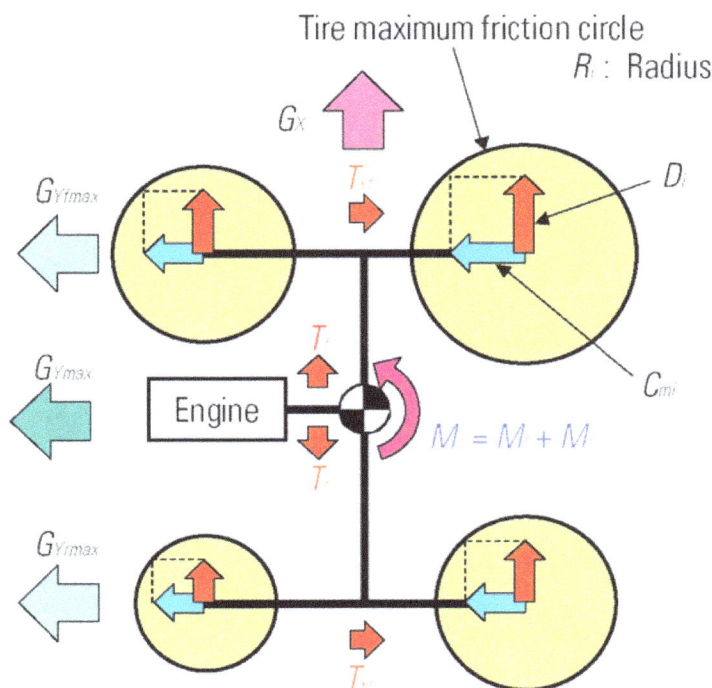

Figure 5. Effect of torque vectoring for four-wheel vehicle [8].

2.8. Active roll control system (ARC)

Active roll control (ARC) is one of the most promising active systems to improve vehicle comfort and operability. When the vehicle is cornering, the vehicle body tilts outward due to centrifugal force, which reduces driving comfort felt by passengers. If the movement of the roll becomes excessive, a fatal rollover may occur. Most passenger cars today are equipped with torsion bars called stabilizers attached to the front and rear axles to reduce roll movement. However, the roll deceleration using the torsion bar also functions as a bridge where the tire vibration on one side is transmitted to the other side. Although the hard stabilizer is good for reducing the roll motion during cornering, it cannot separate vibrations between the left and right tires, which may reduce the comfort during straight driving.

The active roll control system includes a control module, an accelerometer, a speed sensor, a fluid reservoir, an electrohydraulic pump, a pressure control valve, a direction control valve, and a hydraulic actuator on both the front and rear stabilizer bars. Accelerometers and speed sensors may be common to systems other than an active roll control [9]. The configuration of this active roll control system is shown in **Figure 6**. The active roll control system may be referred to as a roll stability control (RSC) system or a dynamic handling system (DHS). Some active roll control systems have gyro sensors that provide voltage signals to the control module in relation to the speed and angle of the vehicle roll. Active roll control can be adopted to improve handling by changing the distribution between the anti-roll torques generated at the front stabilizer bar and the rear stabilizer bar.

When the vehicle is driven straight, the active roll control system does not supply hydraulic pressure to the linear actuator of the stabilizer bar. In this state, both stabilizer bars are free to move until the linear actuator is fully compressed. This action provides improved individual wheel bump performance and better ride comfort. When the chassis begins to tilt during

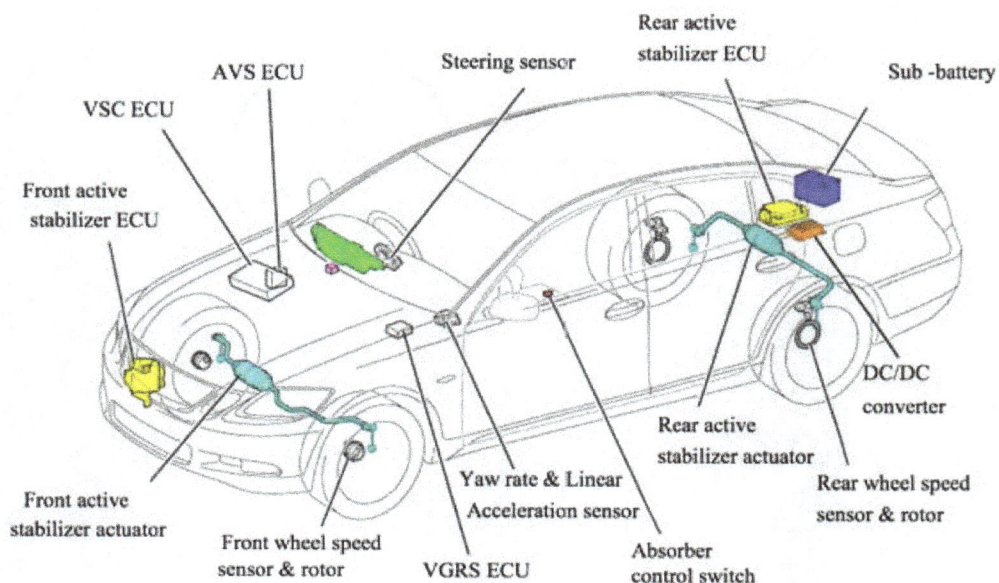

Figure 6. System configuration of active roll control system [9].

cornering, the module operates the directional valve to supply fluid pressure to the linear actuator of the stabilizer bar. This action strengthens the stabilizer bar and reduces the tilt of the body. The active roll control system reduces body lean, enhances safety, and reduces the possibility of rollover of the vehicle.

2.9. Future trends of ABS, TCS, ESC, etc.

As long as the number of built-in system of automotive applications continues to increase, it spreads the possibility of stability of vehicles. Two of the most important advancements affecting the future of ESC are the development of steer-by-wire and brake-by-wire systems. Removal of the mechanical elements from steering and brakes will allow ESC to fully control brake and steering during critical situations. True brake-by-wire is also called EMB or an electromechanical brake. Another form of brake-by-wire is an electrohydraulic brake, but this system uses a conventional method of applying pressure to the brake via a fluid. Full EMB eliminates brake fluid and completely the hydraulic line. The braking force is generated directly on each wheel by a high-performance electric motor. These motors are controlled by the central ECU and are activated by the electrical signals transmitted from the ECU in response to the electronic pedal module. There are not physical connections between the brake pedal and the brake. This mechanical connection is replaced by an electrical signal sent via the bus. Since there is no physical connection to the brake, it is necessary to install an electronic actuator on the brake pedal in order to simulate mechanical feedback to the driver. This is available since several years; ESC in the European Union is mandatory for homologation. The driver cannot control the brake power without braking feedback. Furthermore, since braking is an important safety-related system, in this type of brake-by-wire system, it is necessary to build a fault-tolerant bus or fail-safe system.

A true brake-by-wire vehicle will allow even greater control of the vehicle during system operation. The system will no longer be constrained by the physical limits of the hydraulic brake. There will be little time lag waiting for fluid pressure to build when the system is running. This shortened response time is very precious for hard real-time systems. The second fraction that traditional brakes need to start working means the difference between staying on a frozen road and ending with a groove.

It eliminates the mechanical connection between the steer-by-wire driver and the front tire of the vehicle. Conventional steering elements are replaced by two actuators located at the front corner of the vehicle. These actuators receive input from the control module and turn the front wheels according to the instructions of the control module. The system also uses the electric motor to provide road feedback to the steering wheel.

A true steer-by-wire system offers many advantages over traditional mechanical steering systems. The steer-by-wire has fewer mechanical elements that reduce the vehicle's mass and improve fuel economy in this way. Reducing the number of hardware components can also improve packaging flexibility and system reliability due to the simplification of the vehicle assembly, because system failures are due to component failures. Steer-by-wire also improves the front wheel steering ability which brings a closer turning radius. Steer-by-wire is an embedded system with access to a vehicle network that allows easy integration with other vehicle systems such as ESC. Hand wheel subsystem (steering wheel) includes a torque sensor and a position sensor that measures the force rotating the wheel and the angle that the driver turns the wheels. In addition, there is a motor mechanism that provides feedback to the driver to simulate the traditional "feel" of mechanical steering. The driver needs a torque feedback system to accept the steer-by-wire system. If there is mechanical feedback coming from the steering wheel, the driver might believe that the car will not function properly. The torque feedback device must be steer-by-wire a commercially viable automotive system. It can be categorized as a steer-by-wire hard real-time system. The ability to meet the response time requirements is important for systems such as steer-by-wire. The system must be designed to be fault tolerant because either type of system delay or failure will have disastrous results for the driver.

Steer-by-wire brings many advantages to the vehicle stability system. Steering control eliminates driver interference in critical situations. In a critical situation, the driver may panic and put the vehicle in an uncontrollable state. Although this system is very effective, there is a limit to the ability to stabilize the vehicle. Ultimately, it cannot completely ignore the intention of the driver. Steer-by-wire can further strengthen the control of the vehicle stability system. The ESC has the additional advantage of controlling the orientation and position of the wheel. In the case of steer-by-wire, there is also the possibility to steer each wheel independently as necessary. Steer-by-wire may add other elements to vehicle control, greatly improving ESC performance. On the other hand, for steer-by-wire system manufacturers and vehicle manufacturers, there is a possibility that problems may arise if the car encounters trouble, and then it can result in an accident. When designing and mounting a system like a steer-by-wire, careful consideration must be made so that accidents by the system do not occur.

2.10. Vehicle dynamics computer simulation

The development of automobiles is a very time-consuming and complicated process before being sold to customers after being subjected to various stages of design, development, testing, verification, and improvement. Even using vehicles scaled for testing and development, it takes a lot of time to set up and test the system for each change of vehicle or software. Today, the latest automobiles are equipped with many advanced technological solutions. This, along with strict regulatory requirements for safety and environmental standards, leads to a major challenge in design complexity, cost, and quality. To develop software and hardware used in vehicles, computer simulation is an essential tool for developing and using collision mitigation algorithms in a safe environment, without the capital cost of designing, developing, and setting up test rigs. Testing and verification are very important to ensure the correctness of the function. Simulation is an effective means to reduce time-consuming debugging in the field, allowing more iterations in the design process. Simulation allows you to run thousands of different test scenarios in just a moment of a time to run a full-scale system test. The simulation not only makes it possible to reduce the cost of capital but also visualizes the results from these tests when using graphical simulation and can qualitatively and quantitatively determine the degree of performance of the controller. In addition, it is better to adjust the algorithm parameters as well as to observe the impact of the overall system performance with increments in time and cost and further improve quality.

There are many dynamic car simulators, and it is a very realistic model in terms of vehicle dynamics and driving experience. These simulators and the models have enormous effects to capture the very complex nonlinear situations that affect the dynamic behavior of the vehicle. These simulators can interface with custom vehicle controllers in order to observe the impact on vehicle performance with outstanding reality. Many of these simulators are dedicated to automobile development. Examples include ADAMS/Car, CarSim, CarMaker, and others. Using these simulators, the engineering team can quickly build and test functional virtual prototypes of vehicles and vehicle subsystems, increase productivity, reduce time to market, reduce costs, and improve efficiency. Vehicle dynamics can be investigated due to the simulation of the dynamic behavior of many vehicles including passenger cars, racing cars, light trucks, and utility cars. However, the precision, agility, robustness, efficiency, and intelligence of the motion control system are important for the simulation system. Difficult points for engineers lie in various factors such as nonlinearity, friction, complex internal dynamics, time-varying parameters of system dynamics, working environment disturbances, and complex task tasks. To use advanced control algorithms and schemes, time/frequency domain modeling, system identification, observation of unmeasured state, estimation of important parameters, and corresponding analysis are often necessary [10]. The simulation tools provide a reliable environment in which the tests can be repeated accurately. In addition, simulation and support tools provide an atmosphere to develop a necessary environment because there are test scenarios before installing a driving assistance system on actual vehicles. A simulation tool that performs the hardware in the loop test is used to test and calibrate hardware components such as ECUs by integrating the components in a run test. Since the tool produces very accurate results, the functions of hardware components can be rigorously tested before using them in real vehicles.

3. Conclusion

In this chapter, the means of enhance active and comprehensive safety policy on the automobile dynamics control system is described. The vehicle dynamics control has evolved due to the progress of electronics, miniaturization, and software development and is growing quickly in the future. Moreover, systems of ABS, TCS, ESC, TVC, and ARC are described.

Sensor fusion and connectivity might be more important. Sensor fusion is a technique that combines in multiple physical sensors and generates that combine accurate data even if each sensor alone cannot be trusted. Not only the number and type of sensors but also how to use sensors is important. Software synthesizing results from multiple sources provides more sophisticated analysis with more rapid insight than when data from each sensor must be processed separately. By combining input from various sensors to complement the limits and errors of each sensor, the function can be assured. Automotive suppliers continue to find new ways to replace existing mechanical systems, and it is estimated that in the future, development such fusing and connecting sensors will continue for automatic operations. Electronic vehicle control has been attracting more attention with the attraction of the self-driving vehicles and advanced driver assistance systems.

Today's vehicle network is transforming automobile parts, which are in the areas of mechanical or hydraulic systems or into electronic systems. Replacing mechanical parts with electronic parts will cause an integration of the level of the overall system. As a result, the cost of advanced systems should drop sharply. Sophisticated features such as chassis control and smart sensors are becoming mainstream. The X-by-wire system does not rely on conventional mechanical or hydraulic mechanisms. They make lightweight, cheap, safe, and more fuel-efficient vehicles. Such a system can eliminate a hydraulic brake and pump. The X-by-wire steering system might replace the steering column shaft with an angle sensor and a feedback motor. It will also simplify the production of models for the left hand and the right hand. It is natural to add advanced functions to such electronic systems. In a mechanical steering system, the driver actually feels that the vehicle loses control in an unstable state and can respond appropriately. To accommodate this, the X-by-wire system may include a motor on the steering wheel that provides an artificial feedback to the driver.

Technologies for collecting information with sensors mounted on vehicles have evolved with automatic driving vehicles and advanced driving support systems (ADAS). A system that collects information by various sensors such as cameras that recognize objects and people around vehicles and radars (LIDAR) that detect distant obstacles and oncoming vehicles is being put to practical use. After the surrounding environment can be accurately grasped, a technique is required that accurately utilizes the obtained information for autonomous control of accelerator, brake, and steering wheel. The technology of ABS, TRC, and ESC is applied to the operation of this accelerator, brake, steering wheel.

Author details

Shinji Kajiwara

Address all correspondence to: kajiwara@mech.kindai.ac.jp

Faculty of Science and Engineering, Kindai University, Higashioska, Osaka, Japan

References

[1] Niu S, Zhan W. Analysis of confidence lower limits of reliability and hazard rate for electronic stability control systems. Quality and Reliability Engineering International. 2013;**29**:621-629

[2] Wanner D, Wallmark O, Jonasson M, Drugge L, Stensson A. Control allocation strategies for an electric vehicle with a wheel hub motor failure. International Journal of Vehicle Systems Modelling and Testing. 2015;**10**(3):263-287. DOI: 10.1504/IJVSMT.2015.070164

[3] European Union 2008. Improving the safety and environmental performance of vehicles. Press Release, European Commission—IP/08/786, 23/05/2008

[4] Zhang X, Wei K, Yuan X, Tang Y. Optimal torque distribution for the stability improvement of a four-wheel distributed-driven electric vehicle using coordinated control. Journal of Computational and Nonlinear Dynamics. 2016;**11**(5):051017. DOI: 10.1115/1.4033004

[5] Hart P. ABS Braking Requirements. St Kilda: Hartwood Consulting, Pty Ltd; June 2003

[6] Van Zanten A, Erhardt R, Landesfeind K, Pfaff G. Stability Control. In: Automotive Electronics Handbook. Blacklick: McGraw-Hill; 1999. pp. 17.1-17.33

[7] Liebemann EK, Meder K, Schuh J, Nenninger G. Safety and performance enhancement: The Bosch electronic stability control (ESP). SAE Paper # 05-0471. 2005

[8] Sawase K, Ushiroda Y. Improvement of vehicle dynamics by right-and-left torque vectoring system in various drivetrains. Mitsubishi Motors Technical Review. 2008;**20**:14-20

[9] Buma S, Ookuma Y, Taneda A, Suzuki K, Cho J, Kobayashi M. Design and development of electric active stabilizer suspension system. Journal of System Design and Dynamics. 2010;**4**:61-76. DOI: 10.1299/jsdd.4.61

[10] Li S, Harnefors L, Iwasaki M. Modeling, analysis, and advanced control in motion control systems—Part I. IEEE Transaction on Industrial Electronics. 2016;**63**:5709-5711. DOI: 10.11.09/TIE.2016.2589220

8

Reliable Design of PMa-SynRM

Carlos López-Torres, Antoni Garcia-Espinosa and
Jordi-Roger Riba

Abstract

Electric vehicles require highly reliable and resilient electric motors, due to the harsh operating conditions they must withstand. To this end, there is a current trend to design rare-earth-free machines. Traction electric motors must be optimized in terms of efficiency, torque density, power factor, constant power speed ratio, and cost. Although different technologies are available, permanent magnet assisted synchronous reluctance motors (PMa-SynRM) are promising candidates for such applications. Nowadays, the optimal design process of electrical motors is based on finite element method (FEM) simulations. However, it is very time consuming with a heavy computational burden process, so in order to speed up the optimization process, it is very appealing to have an accurate pre-design of the machine. In this chapter, the electromagnetic pre-design of a PMaSynRM is developed. In the proposed electromagnetic pre-design process, the geometry of the machine is calculated based on analytical equations that take into account the thermal, electrical, magnetic, and mechanical behavior of the machine to ensure a suitable and reliable design.

Keywords: synchronous reluctance machines, pre-design, multi-physics, permanent magnet, electrical machines

1. Introduction

Electric motors for traction applications have been optimized in terms of power density, efficiency, cost, power factor, and wider speed range. There are different motor topologies to meet the requirements of such application, for instance, permanent magnet synchronous motors (PMSM), induction motors (IM), switched reluctance motors (SRM), and synchronous reluctance motors (SynRM). PMSMs have the best power density ratio and can maintain the power in a wide speed range. However, the material cost due to the rare-earth magnets and concerns about availability and supply of the magnets makes it necessary to use other type of motors. In this context, the concept of rare-earth-free-motors gains attention [1]. Synchronous reluctance motors are good candidates in terms of material and manufacture cost; however, the power density and power factor are low. Then, the idea of permanent magnet-assisted synchronous reluctance motors (PMa-SynRM) appears, since they improve the performances of SynRMs by using ferrite magnets. Ferrite magnets have lower electrical conductivity than rare-earth magnets, so the eddy current losses are much lower, and thus the temperature rise due to eddy current losses. Although ferrite magnets exhibit a lower remanent magnetic flux density compared to neodymium magnets, ferrite magnets have a higher Curie temperature. As a consequence, ferrite magnets are well suited to be applied in high-temperature environments, such as electric vehicles, thus offering improved reliability with respect to the use of rare-earth permanent magnets.

The design of a motor usually consists of a multi-physics analysis where the thermal, electric, magnetic, and mechanic domains are analyzed. In the electromagnetic pre-design stage, the geometry of the machine is often calculated based on criteria taking into account different domains. The electromagnetic domain allows calculating the necessary amount of the magnet, the thermal domain settles the size of the slots of the stator, and the mechanical domain settles the size of the radials ribs.

The final geometry of the motor is obtained after an optimization process, where the values of the motor's parameters are variated to determine the best solution. However, the starting point of the design is based on the electromagnetic pre-design. This work aims at guiding the electromagnetic pre-design of the PMa-SynRM.

The pre-design is performed with the basic specifications of the machine, such as mechanical power, corner speed, phase current, pole number, or efficiency required, among others. Since FEA is often not applied to speed up the design process, the parameters calculated must be very accurate, so several refinement loops are introduced. In this context, the whole process is a combination of analytical equations with iterative loops to refine the estimated parameters, which are required to start the electromagnetic pre-design.

The starting point consists of estimating some parameters, such as efficiency, power factor, air gap flux density, or back EMF, in order to determine the required phase current, electrical power, or number of turns per phase, among others. These estimated values depend on the machine type, for instance in the SynRM, the power factor can be selected around 0.7, and the efficiency around 95%.

2. Calculation of electrical parameters

The initial set of equations is given by:

$$P_e = \frac{P_{mec}}{\eta} \tag{1}$$

where "P_{mec}" is the output mechanical power, which is one of the inputs of the electromagnetic pre-design, "η" is the estimated efficiency, and "P_e" is the required electrical power. Using the electrical power calculated in (1), the phase current is given by:

$$I_f = \frac{P_e}{mU_f \cos(\varphi)} \tag{2}$$

Being, "m" and "U_f" are the number of phases and the RMS phase voltage, respectively, which are an inputs, and "$\cos(\varphi)$" is the estimated power factor.

To finalize with the electrical part of the design, the number of turns per phase can be calculated according to (3):

$$N_s = \frac{\sqrt{2}E}{\omega_e k_{w1} l_{eff} \tau_p \frac{D_{is}}{2} \alpha_i \widehat{B}_g} \tag{3}$$

where "E" is the back EMF, which is estimated according to 0.97 of the phase RMS voltage [2]; "ω_e" is the mechanical speed in electrical rad/s of the base point; "l_{eff}" is the effective length of the machine; "$\widehat{B}_g$" is the peak flux density in air gap, which is an estimated value; "k_{w1}" is the winding factor of the fundamental component, which is fixed by the winding distribution; "α_i" is the coefficient to obtain the arithmetical average of the flux density; "D_{is}" is the inner stator diameter; and "τ_p" is the slot pitch.

Considering a sinusoidal flux density, the value of "α_i" is 0.64 ($2/\pi$). However, "α_i" is related in [2].

Note that the effective length and the pole pitch cannot be determined since the air gap volume is unknown. The mains dimensions must be calculated before the number of turns per phase.

3. Main dimensions

The first step to calculate the motor geometry is the determination of the main dimensions of the motor. These parameters are the outer and inner diameter of the rotor, the outer and inner diameter of the stator, and the stack length. It is worthy to mention that depending on the restrictions of the application, the outer dimensions can be fixed. **Figure 1** shows several motor's parameters, such as inner and outer rotor radius (R_{ir} and R_{or}, respectively), inner and outer stator radius (R_{is} and R_{os}, respectively), air gap length (g), slot pitch (τ_s), and pole pitch (τ_p). The stack length (L_{stk}) is the length of the active part, that is, the end winding length is not considered.

The calculation of the motor's geometry starts determining the air gap volume/surface or the outer volume/surface of the machine. In this context, different approaches can be found in the literature to calculate the motor geometry using the data obtained from the specifications. On the one hand, Bianchi et al. [3] and Gamba [4] calculate the exterior geometry, which is represented

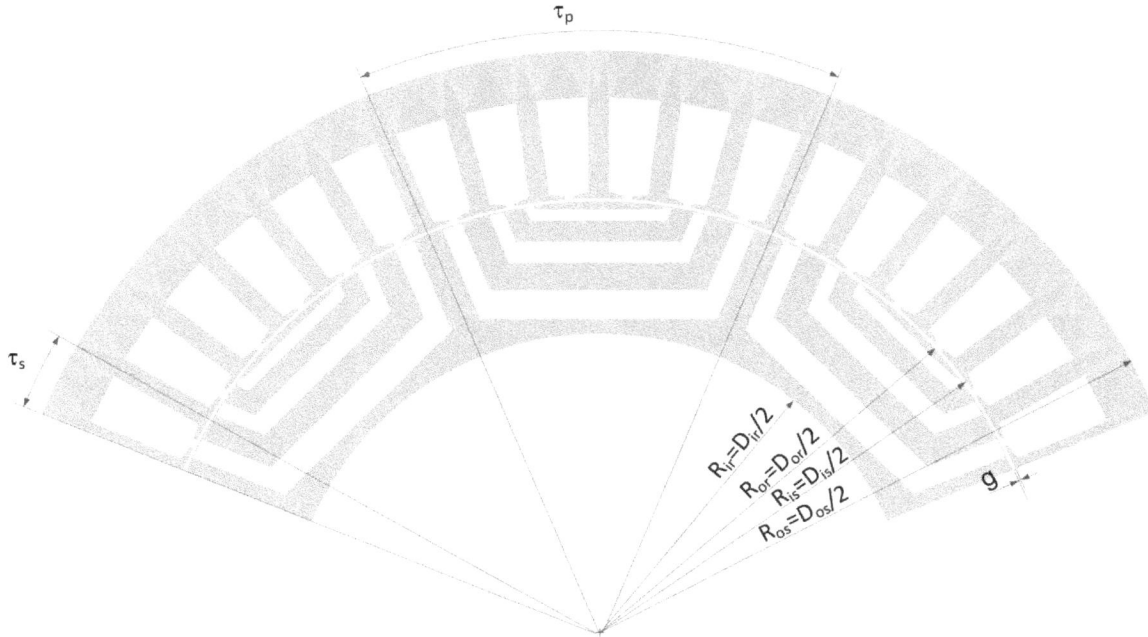

Figure 1. Motor basic geometry.

by the outer stator diameter and the stack length. The first one uses a relation between the torque generated and the volume (K_v), meanwhile the second one relates the losses generated with the outer surface (K_j). According to Bianchi et al. [3], the K_v for these kinds of machines is around 10 Nm/L. However, these values can change depending on the value of the torque. In the second approach, the thermal loading depends on the coolant system, so it is required to determine the outer motor geometry. If the outer geometry is fixed, the thermal loading determines the coolant system required [5]. On the other hand, the electrical loading (A) is used to calculate the geometry of the air gap [6]. In this case, the allowed electrical loading is also determined by the coolant system. Another interesting approach relates the mechanical power of the machine with the air gap volume.

This approach is based on the mechanical constant [2, 7], which is given by:

$$C_{mec} = \frac{P_{mec}}{D_{is}^2 l_{eff} n_{syn}} \tag{4}$$

where "n_{syn}" is the rated electrical frequency.

The proposed pre-design starts with the mechanical constant, so a further explanation of the different values of such constant is required. The value of the mechanical constant is obtained by analyzing several motors of the same typology and coolant systems. **Figure 2**, extracted from [1], shows the relation between the mechanical constant and efficiency for different motor types.

Nevertheless, when using the electrical loading, mechanical constant, or other parameters to obtain the volume or surface of the air gap, the relation between diameter and length is required. The form factor "X" is given by:

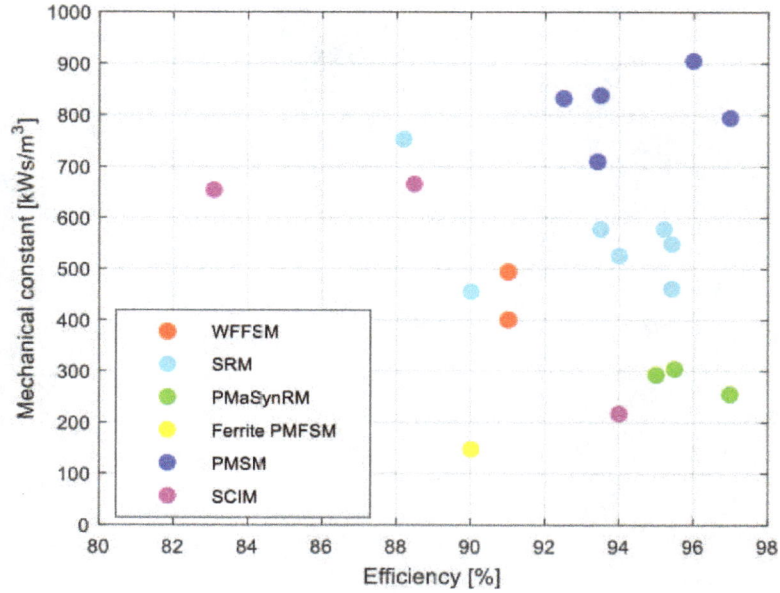

Figure 2. Machines comparison based on the maximum efficiency point and machine constant of mechanical power. Data presented have been collected from [8–24].

$$X \approx \frac{\pi}{4\sqrt{p}} \tag{5}$$

"p" being the number of pole pairs.

Using (5) and (6), the effective length l_{eff} and the bore diameter D_{is} can be calculated. The stack length is obtained according to:

$$L_{stk} = l_{eff} - 2g \tag{6}$$

In order to maximize the saliency ratio, the air gap thickness must be as low as possible [3]. According to Pyrhönen et al. [2], the air gap should be smaller compared to induction machines. The air gap in induction machines is given by:

$$g = \frac{0.18 + 0.006 P_{mec}^{0.4}}{1000} \tag{7}$$

where "P_{mec}" is given in watts.

4. Stator geometry

The stator geometry is completed when the size of the slot, teeth, and yoke are determined. The width of the teeth (b_t) and slots (b_s) can be constant or radial. However, some simplifications can be realized in order to determine the size of these parts. **Figure 3A** shows the geometry of the teeth and slots using the smaller dimension, which it is the most restrictive

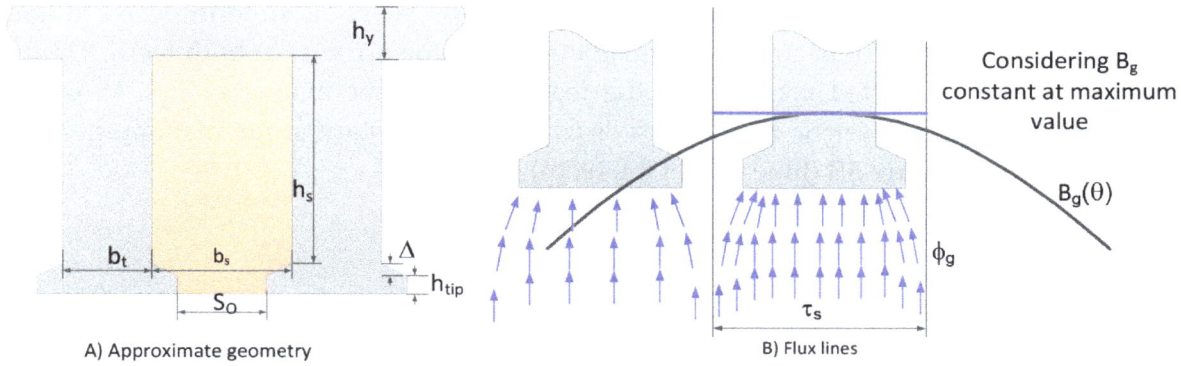

Figure 3. A) Approximate geometry using the most restricted dimensions. B) Flux lines in teeth.

width. The slot opening (S_o) is dimensioned to be higher than the diameter of the conductors. The height of the slot (h_s) and yoke height (h_y) are shown in **Figure 3B**. Finally, the height of tooth tip (h_{tip}) is not important in this stage of the design, although it can be fixed at 1 mm, depending on the machine. In the same way, the "Δ" is not important in the pre-design and can be fixed at 0.5 mm depending on the machine.

In order to calculate the size of the slots, the number of conductors in each slot and the tooth size are required. Then, (3) can be solved since the pole pitch is known. When the number of turns in series per phase is calculated, the number of conductors in each slot (z_q) can be determined as follows:

$$z_q \approx \frac{2mN_{ph}}{Q_s} \tag{8}$$

being, "m" is the number of phases and "a" is the number of parallel paths.

Note that "z_q" must be an integer, so the result obtained needs to be round to obtain the final number of conductors in each slot. Then, the number of turns in series per phase must be updated as:

$$N_{ph\_new} = \frac{z_q Q_s}{2m} \tag{9}$$

In this point, the estimated flux density within the air gap changes, so it is calculated as:

$$\widehat{B}_{g\_new} = \frac{\sqrt{2}E}{\omega_e k_{w1} l_{eff} \tau_p \frac{D_{is}}{2} \alpha_i N_{ph\_new}} \tag{10}$$

Considering a sinusoidal magnetic flux distribution within the air gap and how this magnetic flux distributes through the stator, the size of the teeth and yoke are calculated to obtain a magnetic saturation below a pre-defined value. The teeth have to be sized to withstand the magnetic flux that comes from the air gap. The magnetic flux will enter in the teeth instead of the slots, so the magnetic flux in the pole is divided in the different teeth. However, the teeth are dimensioned considering the worse scenario. **Figure 3B** shows the magnetic fluxes lines of air gap entering in the teeth. The sinusoidal magnetic flux density is superimposed to understand the different quantity or density of magnetic flux (blue arrows) in both teeth. Then, the width of teeth is calculated considering the teeth with higher magnetic flux. In order to oversize the teeth, the magnetic flux density is considered constant at the maximum value as can be observed in **Figure 3B** (blue line in the teeth):

$$\phi_t = \widehat{B}_{g\_new} l_{eff} \tau_s \frac{D_{is}}{2} \tag{11}$$

where "τ_s" is the slot pitch, which is given by:

$$\tau_s = \frac{2\pi}{Q_s} \tag{12}$$

"Q_s" being the number of slots.

In order to obtain a correct size of teeth, the maximum allowed magnetic flux density on this motor part is fixed between 1.5 and 1.3 T. Therefore:

$$b_t = \frac{\phi_t}{k_{sf} \widehat{B}_t} L_{stk} \tag{13}$$

where "$\widehat{B}_t$" is the maximum allowed magnetic flux density and "k_{sf}" is the stacking factor.

On the other hand, the yoke's width must be calculated to drive half of the air gap's magnetic flux on one magnetic pole as can be observed in **Figure 4**.

Then, the magnetic flux in the air gap is calculated as follows:

$$\phi_g = \frac{2}{\pi} \widehat{B}_{g\_new} l_{eff} \tau_p \frac{D_{is}}{2} \tag{14}$$

where the term "$2\widehat{B}_g / \pi$" is used to obtain the average value of a sinusoidal waveform, and "τ_p" is the pole pitch, which are calculated as follows:

$$B_{g\,av} = \frac{1}{\pi} \int_0^\pi \widehat{B}_{g\_new} \sin(\theta) d\theta = \frac{1}{\pi} \widehat{B}_{g\_new}(-\cos(\pi) + \cos(0)) = \frac{2}{\pi} \widehat{B}_g \tag{15}$$

$$\tau_p = \frac{2\pi}{2p} \tag{16}$$

Then, the width of the yoke is:

$$h_y = \frac{\phi_g}{2k_{sf} \widehat{B}_y} L_{stk} \tag{17}$$

where "$\widehat{B}_y$" is the maximum allowed magnetic flux density in yoke, which is fixed between 1.4 and 1.6 T, corresponding to the knee point of the *B-H* curve of the magnetic steel laminations.

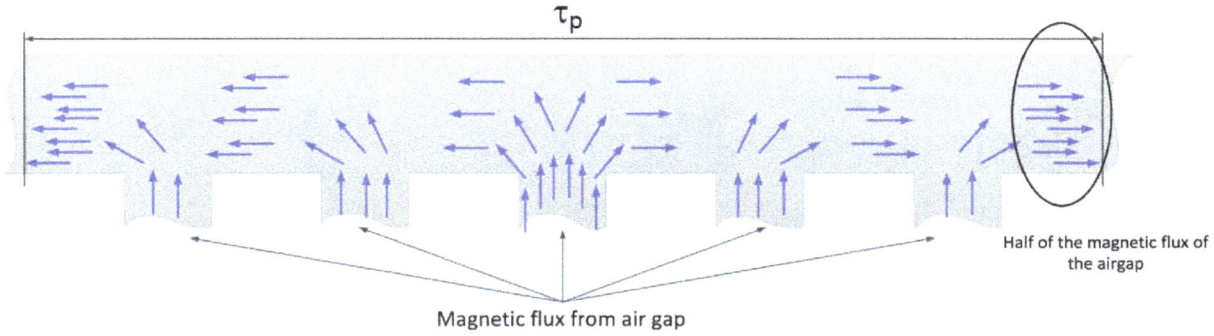

τ_p

Half of the magnetic flux of the airgap

Magnetic flux from air gap

Figure 4. Detail of the magnetic flux in the yoke.

According to **Figure 3A**, when the width of the teeth is known, the slot's width can be determined as:

$$b_s = \tau_s \frac{D_{is}}{2} - b_t \tag{18}$$

Then, height of the slot is determined by:

$$S_{slot} = b_s h_s = \frac{z_q I_{ph}}{J k_u} \rightarrow h_s = \frac{z_q I_{ph}}{J k_u b_s} \tag{19}$$

where "J" is the current density [A/m^2] and variates with the coolant system [2, 6] and "k_u" is the winding factor, which can be fixed at 0.40.

The height of the teeth is given by:

$$h_t = h_s + h_{tip} + \Delta \tag{20}$$

Then, the outer stator diameter is obtained as:

$$D_{os} = D_{is} + 2h_t + 2h_y \tag{21}$$

5. Rotor geometry

The rotor of the SynRMs is punched to create the anisotropy. The insulation, which is the air cavity created in the rotor's perforation, is called flux barrier. The magnetic steel material between flux barriers is called segment or flux carrier. The rotor structure is completed by ribs; there are two different ribs. The first one is the tangential rib, which connects the segments. The other type is the radial rib, which increases the mechanical integrity of the machine. The mentioned parts are depicted in **Figure 5**.

A good saliency ratio can be enhanced by a correct design of the rotor [25]. It starts choosing the proper number of flux barriers [3, 26–28], which is given by:

$$k = \frac{Q_s}{2p} \pm 2 \tag{22}$$

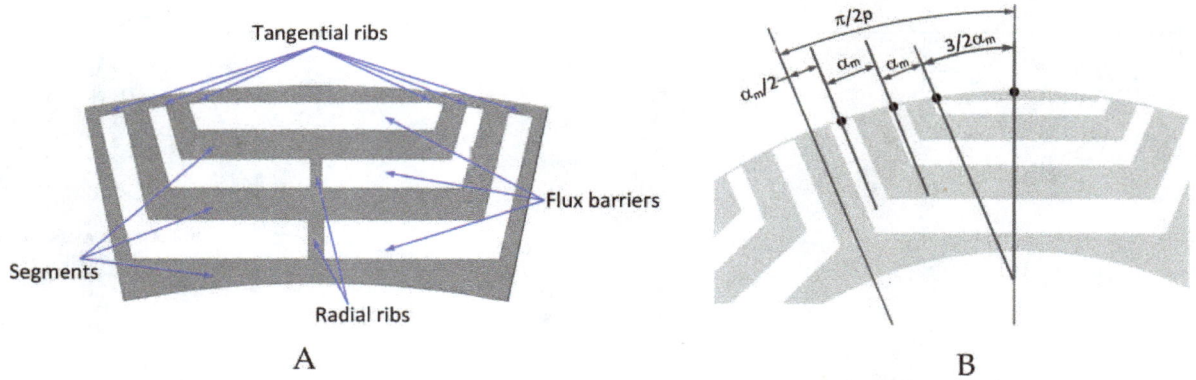

Figure 5. (A) Basic geometry of the rotor of SyrRM. (B) Flux barriers distribution.

Note that according to (22), there are two possibilities for *k*. The choice depends on the application or the rotor size.

Then, the positioning of the barriers is realized to obtain a good distribution of the magnetic flux, that is, a reduction of the torque ripple. The ripple reduction is obtained by means of an optimization process, where the angle of the barriers is changed to find the best solution [29–31]. However, during the pre-design stage, the angle between the end points of the barriers is fixed according to [28, 32]:

$$\alpha_i = \frac{\pi/p}{k+1} \tag{23}$$

Note that (23) calculates the angle between barriers. The angle between the last barrier and the pole center is $3\alpha_m/2$ as can be observed in **Figure 5B**.

The magnetic flux flows through the segments, so a correct sizing is mandatory. Note that the low reluctance of the magnetic steel is related to the magnetic saturation of the segments. Then, the calculation of the width of the flux carriers considers the rotor position with the highest magnetic flux. In this position, which it is called direct axis (d-axis), the maximum magneto-motive force (MMF) in the stator is located between the magnetic poles, meanwhile the zero MMF is in the middle of the magnetic pole, as can be observed in **Figure 6A**. It is worthy to mention that the MMF is considered sinusoidal in order to simplify the calculation of the rotor's geometry.

The widths of the different segments (S_i) are dimensioned to obtain the same magnetic saturation in each segment. In order to estimate the magnetic saturation, the magnetic flux (φ) must be calculated. Considering the geometry shown in **Figure 6A**, an equivalent magnetic circuit can be built to determine the relation between the magnetic fluxes, as depicted in **Figure 7A**. Only one-half of the pole is represented due to the magnetic symmetry.

Note that the reluctance of the air gap is much bigger than the segments' reluctances, so the latter can be disregarded. Then, the magnetic fluxes are given by:

$$\phi_i = \frac{MMF_{di}}{\mathfrak{R}_g} \tag{24}$$

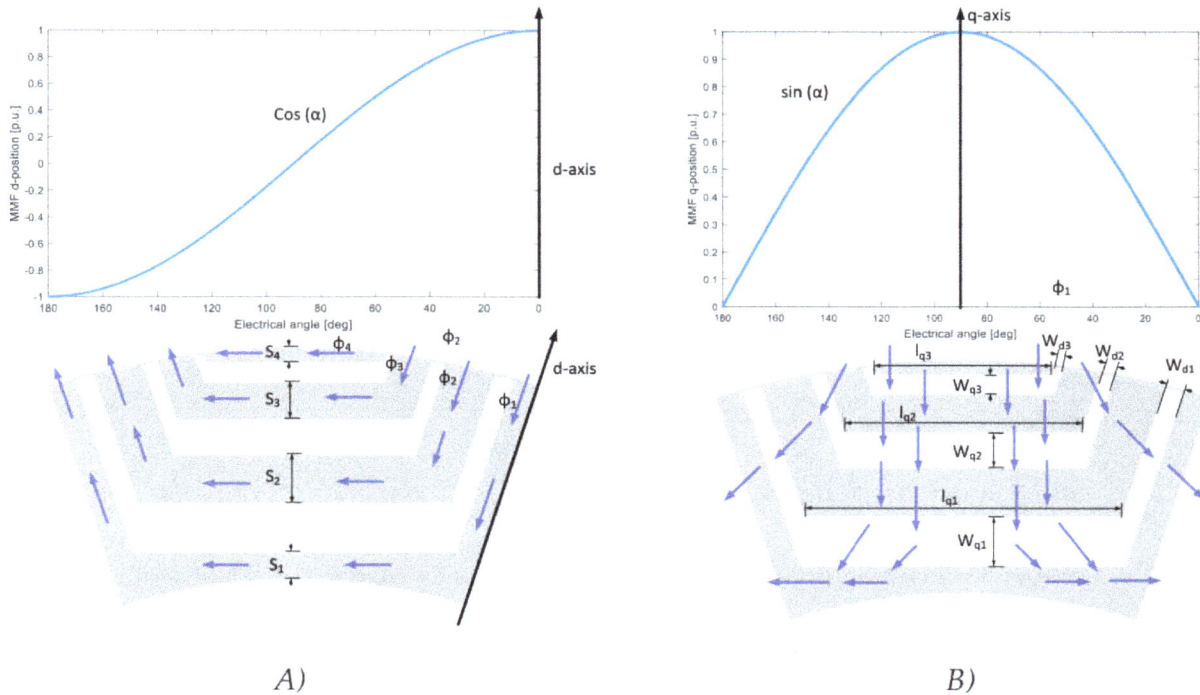

Figure 6. MMF distribution in the dq-positions. The blue arrows represent the magnetic flux in the rotor. (A) d-position (B) q-position.

Therefore, the value of fluxes depends on the MMF. The MMF of each segment is represented by stairs function where the value is the average value of the MMF distribution shown in **Figure 6A**. Since all position angles of the barriers have been fixed, the average MMF of each segment can be calculated as follows:

$$MMF_{di} = \frac{\int_{\frac{2i-3}{2}p\alpha_i}^{\frac{2i-1}{2}p\alpha_i} \cos(\alpha)d\alpha}{p\alpha_i} = \frac{\sin\left(\frac{2i-1}{2}p\alpha_i\right) - \sin\left(\frac{2i-3}{2}p\alpha_i\right)}{p\alpha_i}$$

$$MMF_{dn_b+1} = \frac{\int_{\frac{2n_b-3}{2}p\alpha_i}^{\frac{\pi}{2}} \cos(\alpha)d\alpha}{p\frac{3}{2}\alpha_i} = \frac{1 - \sin\left(\frac{2n_b-3}{2}p\alpha_i\right)}{p\frac{3}{2}\alpha_i}$$

(25)

Then, taking into account (24), where the flux is proportional to the MMF, and the condition of obtaining an equal magnetic saturation on each segment, the relation of segment's width is given by:

$$\frac{S_i}{S_{i+1}} = \frac{MMF_i}{MMF_{i+1}}$$

(26)

A)   B)

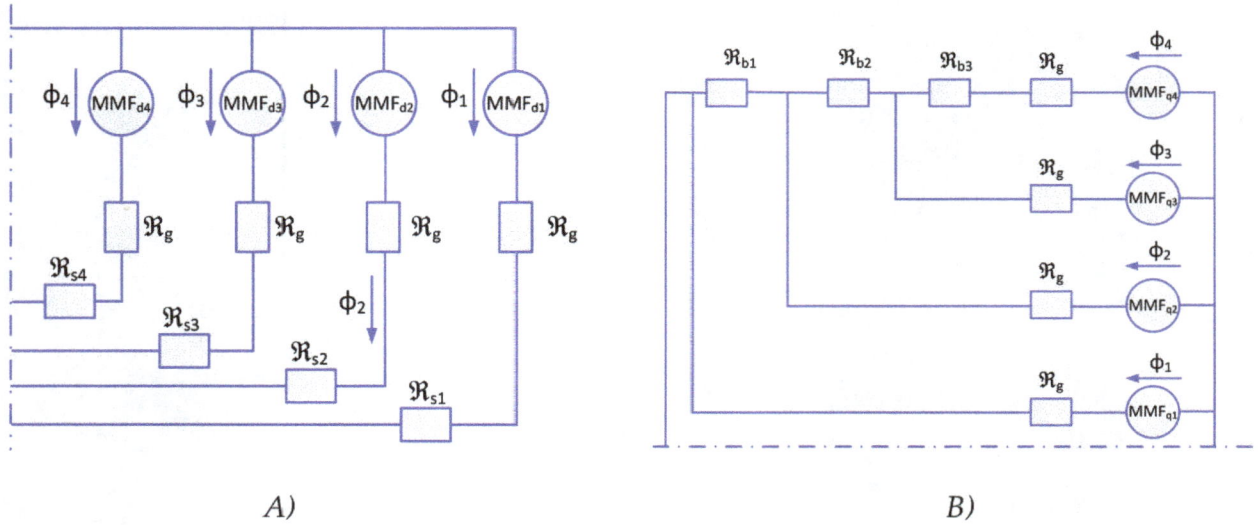

Figure 7. d-Axis equivalent magnetic circuit used to determine the width of the segments. (A) d-position (B) q-position.

The magnetic saturation is calculated using the magnetic flux and the cross section, so (26) is deducted as follows:

$$B_i = \frac{\phi_i}{S_i L_{stk}} \rightarrow B_i = B_{i+1} \rightarrow \frac{\phi_i}{S_i L_{stk}} = \frac{\phi_{i+1}}{S_{i+1} L_{stk}} \rightarrow \frac{\frac{MMF_{di}}{\Re_g}}{S_i L_{stk}} = \frac{\frac{MMF_{di+1}}{\Re_g}}{S_{i+1} L_{stk}} \tag{27}$$

In addition, it is worthy to mention that (26) must be adapted in segment 1, since the magnetic flux is divided in the two magnetic poles, so the final equation to determine the relationship between S_1 and S_2 is given by:

$$\frac{2S_1}{S_2} = \frac{MMF_1}{MMF_2} \tag{28}$$

Finally, there are one more unknowns that equations, so one more equation is required to find out the width of the segments. Since the width of all the segments is equal to the total iron length in the rotor, the last equation results in:

$$L_s = \sum_{i=1}^{i=nb+1} S_i = \frac{h_{rotor}}{1 + k_{insq}} \tag{29}$$

where "h_{rotor}" is calculated as $h_{rotor} = (D_{or} - D_{ir})/2$, and "$k_{insq}$" is the insulation ratio in the q-axis. Note that "k_{insq}", which has values around 1 [2], is defined by:

$$k_{insq} = \frac{L_a}{L_s} \tag{30}$$

"L_a" being the total length of width of air in the rotor given in (37) in the q-axis and "L_s" is the t width of all the segments of magnetic steel along the q-axis.

On the other hand, the flux barriers must be designed to offer a large magnetic resistance to the flow of the magnetic flux. In this context, the sizing of the flux barriers is carried out when the magnetic flux is positioned in the quadrature position, as can be observed in **Figure 6B**. The MMF distribution in the q-position is calculated as in the d-position, that is, using the average value of the MMF considering a sinusoidal distribution, as shown in **Figure 6B**. In this case, the MMF is given by:

$$
MMF_{qi} = \frac{\int_{\frac{2i-3}{2}p\alpha_i}^{\frac{2i-1}{2}p\alpha_i} \sin(\alpha)d\alpha}{p\alpha_i} = \frac{-\cos\left(\frac{2i-1}{2}p\alpha_i\right) + \cos\left(\frac{2i-3}{2}p\alpha_i\right)}{p\alpha_i}
$$

$$
MMF_{qn_b+1} = \frac{\int_{\frac{2n_b-3}{2}p\alpha_i}^{\frac{\pi}{2}} \cos(\alpha)d\alpha}{p\frac{3}{2}\alpha_i} = \frac{\cos\left(\frac{2n_b-3}{2}p\alpha_i\right)}{p\frac{3}{2}\alpha_i}
$$

(31)

Figure 7B depicts the equivalent magnetic circuit in the q-position to calculate the size of the different flux barriers.

Note that the MMF_{q1} is zero, so the path of flux 1 can be removed. As can be observed, the magnetic flux in the q-axis is given by the addition of fluxes 1–4. Then, the purpose of the barriers' sizing is to minimize the q-flux. Then, the relation between the widths of each barrier is given by:

$$
\frac{W_{qi}}{W_{qi+1}} = \frac{MMF_{i+1} - MMF_i}{MMF_{i+2} - MMF_{i+1}} = \frac{\Delta MMF_i}{\Delta MMF_{i+1}}
$$

(32)

A demonstration of the procedure to obtain (32) is further developed. However, the following example is realized with two barriers for the sake of simplification. **Figure 8** shows a rotor with two flux barriers, the magnetomotive force in each segment, which has been calculated with (31), and the variable to optimize, which is the size of the first barrier.

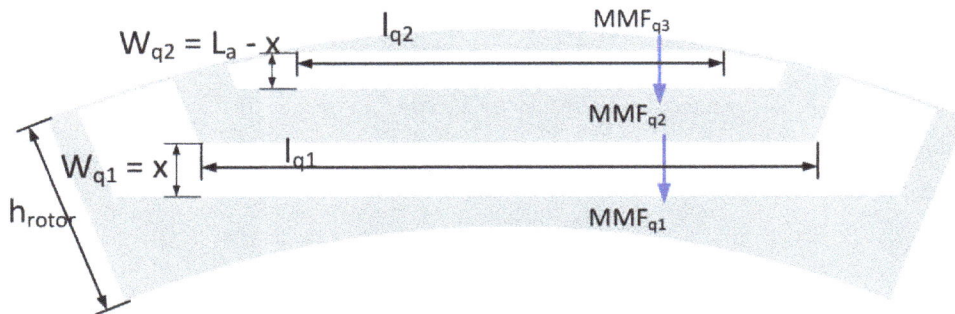

Figure 8. Example of two barriers to determine the relation between the widths of the barriers.

As mentioned before, the sizing of the barriers aims to reduce the total flux in the q-axis (blue arrows). Then, the flux in each barrier is given by:

$$\phi_i = \frac{MMF_{qi+1} - MMF_{qi}}{\mathfrak{R}_{bi}} = \frac{\Delta MMF_i}{\mathfrak{R}_{bi}} \tag{33}$$

The reluctance of the barriers is given by:

$$\mathfrak{R}_{bi} = \frac{W_{qi}}{\mu_o l_{qi} L_{stk}} \tag{34}$$

Therefore, the total flux is given by:

$$\phi = \frac{\Delta MMF_1}{W_{q1}} \mu_o l_{q1} L_{stk} + \frac{\Delta MMF_2}{W_{q2}} \mu_o l_{q2} L_{stk} \tag{35}$$

Note that the total flux is a function of the reluctances of the flux barriers, that is, the total flux is a function of the variable "x" (see **Figure 8**). Hence:

$$\phi(x) = \frac{\Delta MMF_1}{x} \mu_o l_{q1} L_{stk} + \frac{\Delta MMF_2}{L_a - x} \mu_o l_{q2} L_{stk} \tag{36}$$

where the total width of air in the rotor is given by:

$$L_a = \sum_{i=1}^{i=nb} W_{qi} = \frac{h_{rotor}}{1 + \frac{1}{k_{insq}}} \tag{37}$$

Then, the minimization of the flux is obtained as follows:

$$\frac{d\phi(x)}{dt} = -\frac{\Delta MMF_1}{x^2} \mu_o l_{q1} L_{stk} + \frac{\Delta MMF_2}{(L_a - x)^2} \mu_o l_{q2} L_{stk} = 0 \tag{38}$$

Then, the final result is:

$$\frac{\Delta MMF_1}{x^2} l_{q1} = \frac{\Delta MMF_2}{(L_a - x)^2} l_{q2} \rightarrow \frac{\Delta MMF_1}{W_{q1}^2} l_{q1} = \frac{\Delta MMF_2}{W_{q2}^2} l_{q2} \tag{39}$$

Finally, the permeance of each barrier can be assumed constant in order to obtain a better distribution of the flux in the air gap:

$$\frac{l_{q1}}{W_{q1}} = \frac{l_{q2}}{W_{q2}} \tag{40}$$

Finally, by introducing (40) in (39), (32) appears.

It is worthy to mention that there is another approach [27, 28], which relates the size of the barriers as follows:

$$\frac{W_{qi}}{W_{qi+1}} = \frac{\Delta MMF_i^2}{\Delta MMF_{i+1}^2} \tag{41}$$

As can be observed in **Figure 6B**, there is another variable to define, which it is the width of the flux barrier in the lateral location (W_{di}). The relation between the widths of the barriers in the q-axis with the thickness in the d-axis is given by [28]:

$$\frac{W_{di}}{W_{di+1}} = \frac{W_{qi}}{W_{qi+1}} \tag{42}$$

Note that one more equation is required to solve the sizing of the barriers. The total length of the barriers can be determined by using the insulation ratio in the d-axis:

$$L_{ad} = \sum_{i=1}^{i=nb} W_{di} = L_s k_{insd} \tag{43}$$

"k_{insd}" being the insulation ratio in the d-axis, which is applied to determine the width of the barriers according to the mechanical angle defined in **Figure 5B**. "L_s" is the total thickness of the segments, which it is constant in the whole segment.

In this point, the sizing of the rotor is explained. However, there are several uncertain points. These undefined variables are the inner rotor diameter and the insulation ratios (k_{insd} and k_{insq}).

On the one hand, the inner rotor diameter (D_{ir}) determines the total space in the rotor, since the outer rotor diameter is known:

$$D_{or} = D_{is} - 2g \tag{44}$$

Then, the inner rotor diameter is defined as:

$$D_{ir} = D_{or} - 2h_{rotor} \tag{45}$$

"h_{rotor}" is required to calculate the width of the barriers and segments, as can be observed in (29) and (37). Then, an iterative system to determine the correct size of the rotor is proposed as can be observed in **Figure 9**. Depending on the design restrictions, this part must be adapted. For instance, the use of magnets makes necessary to size the barriers with thickness greater than a certain value, which depends on the magnet's manufacturer (around 3 mm), or to increase the rotor size to obtain a desired saliency ratio, or to introduce the necessary quantity of magnet to improve the motor performances in terms of constant power ratio. In this context, **Figure 9** shows the iterative design procedure. Furthermore, an example of the differences is depicted. In the first iteration (i = 1), the width of the last barrier is 0.5 mm, however, the specifications only allow values higher than 3 mm. Then, after the iterative procedure (i = n), the solution is obtained according to the restrictions imposed. Note that the inner diameter of

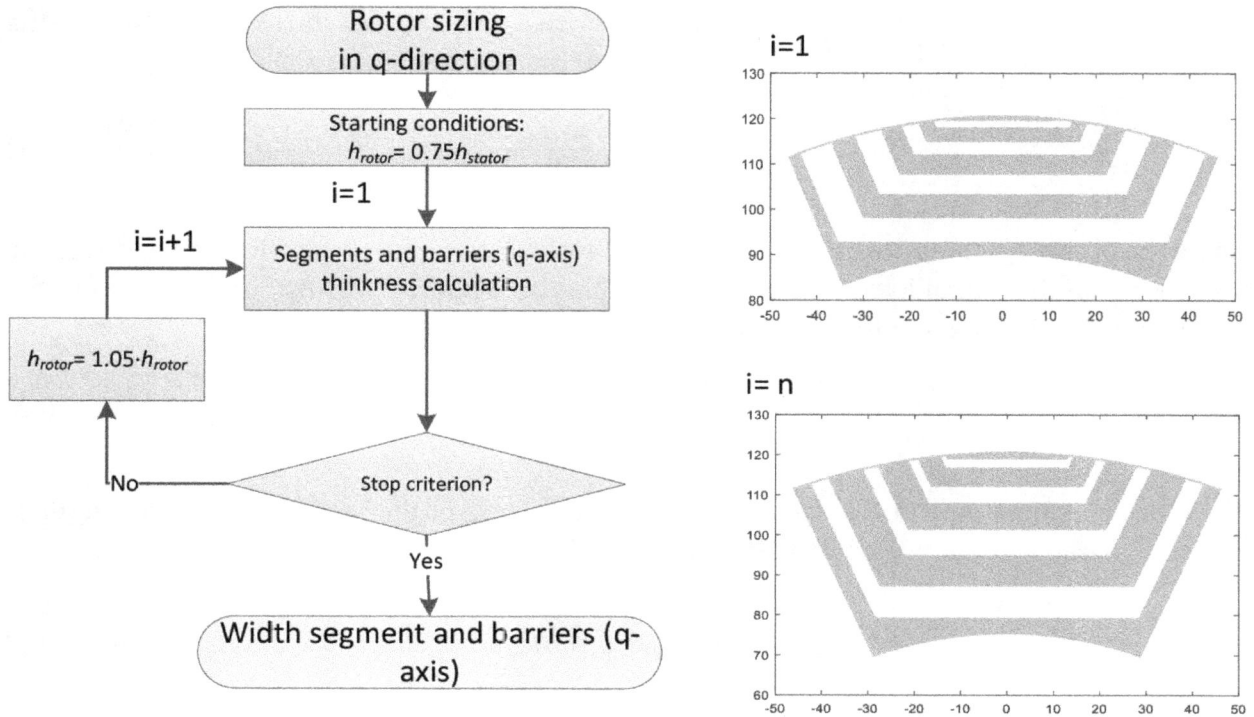

Figure 9. Iterative loop to size the rotor segments and barriers in the q-position.

the rotor in the first case is 90 mm meanwhile in the final solution is 75 mm. The insulation ratio in the q-axis is defined by the designer, and probably the best option is obtained after an optimization. However, values around 1 are good solutions.

On the other hand, the d-axis insulation ratio is not defined. As explained before, this insulation ratio is determined to locate the barriers according to **Figure 5**. Then, the value of this variable is swept to obtain the final design. In this case, the criterion to halt the iterative process is the correct position of the last barrier (the angle is $1.5\alpha_m$). **Figure 10** shows the iterative procedure and the solutions of two different iterations.

It is noted that a posterior mechanical verification is required to ensure a suitable mechanical strength of the rotor configuration obtained in this step.

After the calculation of the rotor size, the magnet quantity must be determined in order to obtain a suitable behavior of the machine during the operation. The north of the magnet is located in the negative direction of the q-axis (see **Figure 11**) in order to improve the motor capabilities, such as torque, base speed, and angle between voltage and current (see **Figure 11B**).

The motor capability within the flux-weakening region is related with the magnet contribution [33, 34]. It means that the magnets can be or not be inserted in all the barriers, depending on the requirements. In the case of not inserting magnets in all barriers, it is recommended to put the magnets in the innermost barrier, since the outset barriers are more magnetically stressed, so the magnet can suffer demagnetization [3, 35].

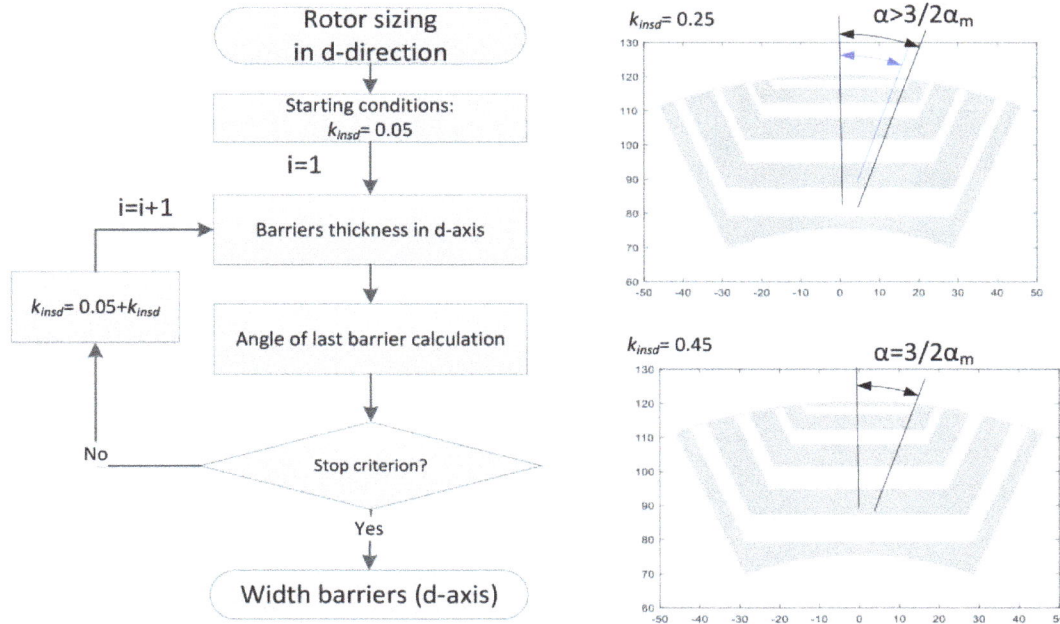

Figure 10. Rotor iterative loop to size the segments and barriers in d-position.

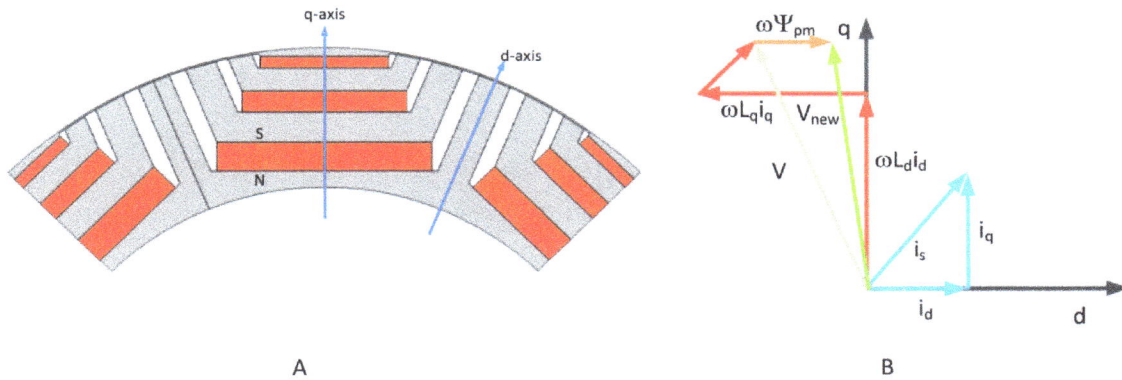

Figure 11. (A) Magnet orientation in the rotor. (B) Phasor diagram with and without magnets.

To compute the inductances and the magnet flux linkage, a fast and simple magnetic model is introduced. It is important to mention that a complex magnetic model is required in the optimization stage [36, 37]; however, in the electro-magnetic pre-design, the proposed magnetic model is good enough to calculate the magnet and improve the accuracy of the geometry.

The magnetic model based on two reluctance networks (RN) not only calculates the magnetic flux linkage but also estimates the dq-inductances, so the motor performances can be calculated. Using this magnetic model, the back EMF, power factor, efficiency, saturation factor, and air gap flux density can be defined with more accuracy and the electromagnetic pre-design can be improved (see Algorithm 1).

There are two reluctance networks to analyze, the d- and q-axis. The q-reluctance network (RN) allows calculating the q-inductances and the magnet flux linkage. **Figure 12** shows an equivalent magnetic model to determine the mentioned parameters of the machine.

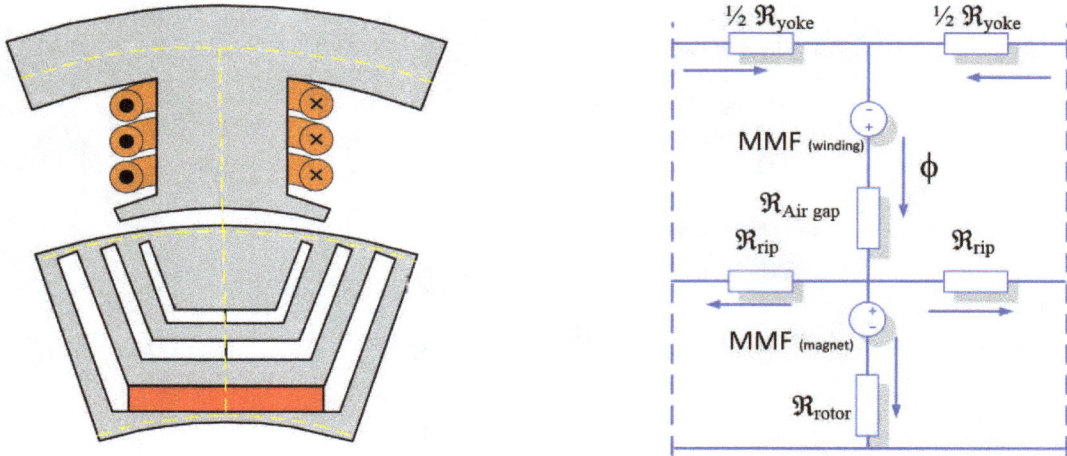

Figure 12. Simple reluctance network in q-axis.

Note that, the magnet is only located in the innermost barrier, however, it can be removed or more magnets can be introduced in the remaining barriers. The RN is formed by MMF generators and reluctances. The first one is created by the coils and magnets:

$$MMF_{winding} = \sum NI \tag{46}$$

"N" being the number of conductors in the coil and "I" the current in each phase (in this, select the rated current).

It is worthy to be mentioned that this magnetic model is not complete. The winding MMF generator is only represented in one tooth, so the whole contribution of the different teeth has to be added. In [38], there is more information to calculate the MMF$_{winding}$ according to a given winding distribution:

$$MMF_{magnet} = H_c W_{qi} \tag{47}$$

where "H_c" is the coercive force and "W_{qi}" is the width of the magnet, which is the same width as that of the flux barriers. Then, the reluctances are calculated as follows:

$$\Re = \frac{l}{\mu_o \mu_r S} \tag{48}$$

where "l" is the length of the magnetic reluctance, "S" is the cross section of the magnetic reluctance, "μ_o" is the magnetic permeability of the vacuum, and "μ_r" is the relative magnetic permeability, which in the air is equal to 1 meanwhile in the magnetic steel varies with the saturation. Note that the magnetic steel is not saturated along the q-axis. Finally, the air gap reluctance must be multiplied by Carter's coefficient to reflect the effect of the slot opening.

On the other hand, the d-axis RN is shown in **Figure 13**. In this case, the magnet is not reflected since it only influences the q-axis. The magnetic saturation of the magnetic steel must be considered. The magnetic saturation in the teeth and yoke can be fixed at the value chosen in the design stage (13) and (17), and in the rotor can be fixed at 1 T.

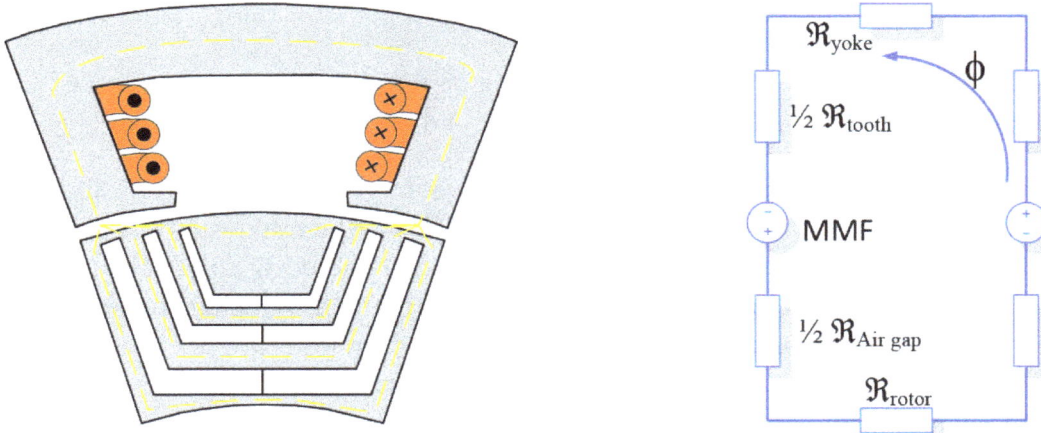

Figure 13. Simple reluctance network in d-axis.

The unknown magnetic flux is calculated using two different equations. The first one (49) relates the MMF obtained in a closed path with the reluctances and the magnetic flux in this path:

$$MMF = \sum \Re \phi \tag{49}$$

The second one relates the total magnetic fluxes in a node (see **Figure 12**):

$$\sum \phi = 0 \tag{50}$$

The q-axis is solved twice; the first one only considers the MMF generated by the magnets in order to calculate the magnet flux linkage (Ψ_{mpq}), meanwhile the second one takes into account both MMF generators. The inductances and flux magnet linkage are calculated as follows:

$$L_d = 2pN\frac{\phi_d}{I}; L_q = \frac{2pN\phi_q - \Psi_{mpq}}{I}; \Psi_{mpq} = 2pN\phi_q \; where \; I_q = 0 \tag{51}$$

By using these values, the motor performances can be deduced, so the process could be restarted with these new values. The back EMF is calculated as follows:

$$E_d = -\omega_e L_q i_q - \omega_e \Psi_{mpq}; E_q = \omega_e L_d i_d \tag{52}$$

The current angle is given according to the MTPA rule, so the d- and q-currents are known. Then, the power factor can be deduced by calculating the phase shift between the current and voltage. Finally, the torque and output power can be calculated:

$$T = \frac{m}{2}p\left((L_d - L_q)i_d i_q - \Psi_{mpq}i_d\right) \tag{53}$$

$$P_{out} = \omega_e T \tag{54}$$

Then, the losses, which are composed by copper and iron losses, are given by:

$$P_{cu} = mR_s I_{rms}^2 \tag{55}$$

$$P_{fe} = k_h \frac{\omega_e}{2} \widehat{B}^{n_i} + k_e \left(\frac{\omega_e}{2} \widehat{B}\right)^2 \tag{56}$$

"R_s" being the phase resistance calculated with the geometry obtained during the calculation of the stator geometry, "m" the number of phases, and "I_{rms}" the rated current in rms value. On the other hand, the iron losses are composed by hysteresis and eddy current components. The hysteresis and eddy current coefficients (k_h, k_e, n_i) are obtained by using the material specific losses obtained from the manufacturer's datasheets. The iron loses are per unit of mass, so the final value must be multiplied by the mass of the different parts. The computed parts can be the yoke, teeth, and rotor, since these three parts have different magnetic saturation. Then, the efficiency is given by:

$$\eta = \frac{P_{out}}{P_{out} + P_{cu} + P_{fe}} \tag{57}$$

In this point, the thermal behavior of the machine has been considered in the sizing of the slot. The magnetic behavior is analyzed by using the proposed simple magnetic model. Then, the oversize of the slots and the magnet compensated situation ensure the reliability of the motor in terms of magnetic and thermal behaviors. However, the mechanical stress has to be considered to ensure the correct behavior of the machine, since the rotor structure reduces the mechanical integrity.

The mechanical problems are solved by the correct sizing of the radial ribs. Several authors deal with this problem [39–41]. The centrifugal force is given according to:

$$F_c = M\omega_m^2 R_G \tag{58}$$

where "M" is the mass that the calculated radial rib has to support, "R_G" is the radius of the gravity center of the mass, and "ω_m" is the mechanical speed.

Then, the width of the radial ribs (W_r) is given as:

$$W_r = \frac{k_s F_c}{\sigma_r L_{stk}} \tag{59}$$

where "k_s" is the safety factor, which is chosen over 2 and "σ_r" is the tensile strength of the lamination.

6. Summary of the design procedure

In this point, the whole process to obtain the electromagnetic pre-design according to the given requirements is realized. A summary detailing the parameters and equations required in each step is shown in Algorithm 1.

1: Introduce the desired performances (power, rated speed)

2: Introduce the fixed parameters (pole pairs, phase number, slots, Bus DC)

3: Start electro-magnetic pre-design process

4: Estimate parameters (efficiency, power factor, back EMF, saturation factor)

5: **while** stop criterion is not achieved **do**

6:     Basic parameters calculation: Electric power (1), phase current (2)

7:     Estimate C_{mec} according to **Figure 2**

8:     Calculate D_{is} and l_{eff} using (4) and (5). Compute the g (7) and L_{stk} (6)

9:     Estimate number of turns in series N_{ph} (3)

10:     Calculates the number of conductors in slot (8)

11:     Calculates the number of turns in series N_{ph} new (9) and the B_g (10)

12:     Calculates the stator geometry (teeth, yoke, and slots dimensions) (11–21)

13:     Chose the number of flux barriers (22)

14:     Calculates the position of the barriers (23)

Calculation of rotor in q-axis (**Figure 9**)

15:         **while** stop criterion is not achieved **do**

16:             Define h_{rotor}

17:             Calculates the width of the segments and barriers (25–30) and (31–32)

18:             Evaluates stop criterion

19:         **end**

Calculation of rotor in d-axis

(**Figure 10**)

20:         **while** stop criterion is not achieved **do**

21:             Define k_{insd}

22:             Calculates the width of the barriers in d-axis(42–43)

23:             Evaluates stop criterion

24:         **end**

25:         *Solve Magnetic model* (**Figure 12** *and* **Figure 13**)

26:         Calculates Inductances, magnetic flux linkage (51)

27:         Calculates losses (55–56)

28:         Calculates motor performances: Torque (53), output power (54)

29:         Calculates the estimated values: Back EMF (52), efficiency (57), power factor, saturation

factor, peak air gap flux density (using the d-flux from magnetic model).

30:         Evaluates stop criterion (error of estimated parameters)

31:**end**

32: Calculates the thickness of the radial ribs (58–59)

7. Conclusions

Due to the harsh operating conditions, electric vehicles require highly reliable and resilient electric motors. To this end, rare-earth-free PMaSynRMs are appealing candidates. In this chapter, a design procedure of PMaSynRMs has been presented, which includes electromagnetic, thermal, and mechanical restrictions in order to ensure a reliable and resilient operation within extended operational limits. For example, in the event of a major demagnetization failure, the PMaSynRM designed following the proposed approach is able to work as a synchronous reluctance machine, thus providing about 75% of the rated torque. In addition, the use of ferrite magnets allows the machine to operate in higher temperature environments.

Author details

Carlos López-Torres*, Antoni Garcia-Espinosa and Jordi-Roger Riba

*Address all correspondence to: carlos.lopez.torres@upc.edu

Electrical Engineering, Universitat Politècnica de Catalunya (UPC), Terrassa, Spain

References

[1] Riba J-R, López-Torres C, Romeral L, Garcia A. Rare-earth-free propulsion motors for electric vehicles: A technology review. Renewable and Sustainable Energy Reviews. 2016;**57**:367-379

[2] Pyrhönen J, Jokinen T, Hrabovková V. Design of Rotating Electrical Machines. second ed. Wiley; 2014

[3] Bianchi N, Mahmoud H, Bolognani S. Fast synthesis of permanent magnet assisted synchronous reluctance motors. IET Electric Power Applications. 2016;**10**(5):312-318

[4] M. Gamba, "Design of Non Conventional Synchronous Reluctance Machine," Politecnico di Torino, 2017

[5] Lu C, Ferrari S, Pellegrino G. Two design procedures for PM synchronous Machines for Electric Powertrains. IEEE Transactions on Transportation Electrification. 2017;**3**(1): 98-107

[6] Hendershot JR a M. T.J.E., Design of Brushless Permanent-Magnet Motors (Monographs in Electrical and Electronic Engineering). Hillsboro, OH : Oxford: Magna Physics Pub: Motor Design Books LLC; 1994

[7] Huppunen J. High-Speed Solid-Rotor Induction Machine - Electromagnetic Calculation and Design. Finland: Lappeenranta University of Technology; 2004

[8] Dorrell DG, Knight AM, Popescu M, Evans L, Staton DA. Comparison of different motor design drives for hybrid electric vehicles. In: 2010 IEEE Energy Conversion Congress and Exposition. 2010. pp. 3352-3359

[9] Petrov I, Pyrhonen J. Performance of low-cost permanent magnet material in PM synchronous machines. Industrial Electronics, IEEE Transactions on. 2013;**60**(6):2131-2138

[10] Montalvo-Ortiz EE, Foster SN, Cintron-Rivera JG, Strangas EG. Comparison between a spoke-type PMSM and a PMASynRM using ferrite magnets. In: Electric Machines & Drives Conference (IEMDC), 2013 IEEE International; 2013. pp. 1080-1087

[11] Chiba A, Takeno M, Hoshi N, Takemoto M, Ogasawara S, Rahman MA. Consideration of

number of series turns in switched-reluctance traction motor competitive to HEV IPMSM. IEEE Transactions on Industry Applications. 2012;**48**(6):2333-2340

[12] Morimoto S, Ooi S, Inoue Y, Sanada M. Experimental evaluation of a rare-earth-free PMASynRM with ferrite magnets for automotive applications. IEEE Transactions on Industrial Electronics. 2014;**61**(10):5749-5756

[13] Chiba K, Chino S, Takemoto M, Ogasawara S. Fundamental analysis for a ferrite permanent magnet axial gap motor with coreless rotor structure. In: Electrical Machines and Systems (ICEMS), 2012 15th International Conference on, 2012. pp. 1-6

[14] Sone K, Takemoto M, Ogasawara S, Takezaki K, Akiyama H. A ferrite PM in-wheel motor without rare earth materials for Electric City commuters. IEEE Transactions on Magnetics. 2012;**48**(11):2961-2964

[15] Miura T, Chino S, Takemoto M, Ogasawara S, Akira C, Nobukazu H. A ferrite permanent magnet axial gap motor with segmented rotor structure for the next generation hybrid vehicle. In: Electrical Machines (ICEM), 2010 XIX International Conference on, 2010. pp. 1-6

[16] Chiba A, Kiyota K, Hoshi N, Takemoto M, Ogasawara S. Development of a rare-earth-free SR motor with high torque density for hybrid vehicles. IEEE Transactions on Energy Conversion. 2015;**30**(1):175-182

[17] Obata M, Morimoto S, Sanada M, Inoue Y. Performance evaluation of high power and low torque ripple structure of rare-earth free PMASynRM with ferrite magnet. In: Power Electronics and Drive Systems (PEDS), 2013 IEEE 10th International Conference on, 2013. pp. 714-719

[18] Obata M, Morimoto S, Sanada M, Inoue Y. Performance of PMASynRM with ferrite magnets for EV/HEV applications considering productivity. Industry Applications, IEEE Transactions on. 2014;**50**(4):2427-2435

[19] Pellegrino G, Vagati A, Boazzo B, Guglielmi P. Comparison of induction and PM synchronous motor drives for EV application including design examples. IEEE Transactions on Industry Applications. 2012;**48**(6):2322-2332

[20] Dorrell DG, Knight AM, Evans L, Popescu M. Analysis and design techniques applied to hybrid vehicle drive machines—Assessment of alternative IPM and induction motor topologies. Industrial Electronics, IEEE Transactions on. 2012;**59**(10):3690-3699

[21] Mishra P, Saha S. Design modeling and simulation of low voltage squirrel cage induction motor for medium weight electric vehicle. In: Advances in Computing, Communications and Informatics (ICACCI), 2013 International Conference on, 2013. pp. 1697-1704

[22] Chino S, Ogasawara S, Miura T, Chiba A, Takemoto M, Hoshi N. Fundamental characteristics of a ferrite permanent magnet axial gap motor with segmented rotor structure for the hybrid electric vehicle. In: 2011 IEEE Energy Conversion Congress and Exposition. 2011. pp. 2805-2811

[23] Liu Y, Zhao J, Wang R, Huang C. Performance improvement of induction motor current controllers in field-weakening region for electric vehicles. IEEE Transactions on Power Electronics. 2013;**28**(5):2468-2482

[24] Kiyota K, Kakishima T, Chiba A. Comparison of test result and design stage prediction of switched reluctance motor competitive with 60-kW rare-earth PM motor. Industrial Electronics, IEEE Transactions on. 2014;**61**(10):5712-5721

[25] Niazi P, Toliyat HA, Dal-Ho C, Jung-Chul K. A low-cost and efficient permanent-magnet-assisted synchronous reluctance motor drive. Industry Applications, IEEE Transactions on. 2007;**43**(2):542-550

[26] Niazi P, Toliyat HA. Permanent magnet assisted synchronous reluctance motor, design and performance improvement, College Station, Tex.: Texas A&M University, 2006. [Online]. Available: http://hdl.handle.net/1969.1/3178

[27] Vagati A, Pastorelli M, Francheschini G, Petrache SC. Design of low-torque-ripple synchronous reluctance motors. Industry Applications, IEEE Transactions on. 1998;**34**(4):758-765

[28] Moghaddam RR, Gyllensten F. Novel high-performance SynRM design method: An easy approach for a complicated rotor topology. IEEE Transactions on Industrial Electronics. 2014;**61**(9):5058-5065

[29] Pellegrino G, Cupertino F, Gerada C. Automatic Design of Synchronous Reluctance Motors Focusing on barrier shape optimization. Industry Applications, IEEE Transactions on, 2015;**51**(2):1465-1474

[30] Ferrari M, Bianchi N, Doria A, Fornasiero E. Design of synchronous reluctance motor for hybrid electric vehicles. In: Electric Machines & Drives Conference (IEMDC), 2013 IEEE International. 2013. pp. 1058-1065

[31] Barcaro M, Pradella T, Furlan I. Low-torque ripple design of a ferrite-assisted synchronous reluctance motor. IET Electric Power Applications. 2016;**10**(5):319-329

[32] Khan KS. Design of a Permanent-Magnet Assisted Synchronous Reluctance Machine for a Plug-In Hybrid Electric Vehicle. School of Electrical Engineering, KTH, Stockholm, Sweden: Electrical Machines and Power Electronics; 2011

[33] Carvajal Almendros C. Design and Analysis of a Fractional-Slot Concentrated-Wound Permanent-Magnet-Assisted Synchronous Reluctance Machine. Electrical Engineering, KTH Royal Institute of Technolog, Stockholm, Sweeden: Electronic Power Engineering; 2015

[34] Wang Y, Bacco G, Bianchi N. Geometry analysis and optimization of PM-assisted reluctance motors. IEEE Transactions on Industry Applications. 2017;**53**(5):4338-4347

[35] Bianchi N, Fornasiero E, Carraro E, Bolognani S, Castiello M. Electric vehicle traction based on a PM assisted synchronous reluctance motor. In: Electric Vehicle Conference (IEVC), 2014 IEEE International. 2014. pp. 1-6

[36] Torres CL, Michalski T, Espinosa AG, Romeral L. Fast optimization of the magnetic model by means of reluctance network for PMa-SynRM. In: IECON 2016 - 42nd Annual Conference of the IEEE Industrial Electronics. Society. 2016. pp. 1642-1647

[37] Torres CL, Garcia A, Riba J, Romeral L. Design and optimization for vehicle driving cycle of rare-earth-free SynRM based on coupled lumped thermal and magnetic networks. IEEE Transactions on Vehicular Technology. 2017;**99**:1-1

[38] Lopez Torres C, Michalski T, Garcia A, Romeral L. Rotor of Synchronous Reluctance Motor optimization by means reluctance network and genetic algorithm. ed. Electrical Machines and Systems (ICEMS), 2016

[39] Barcaro M, Meneghetti G, Bianchi N. Structural analysis of the interior PM rotor considering both static and fatigue loading. IEEE Transactions on Industry Applications. 2014; **50**(1):253-260

[40] Cupertino F, Palmieri M, Pellegrino G. Design of high-speed synchronous reluctance machines. In: 2015 IEEE Energy Conversion Congress and Exposition (ECCE). 2015. pp. 4828-4834

[41] Babetto C, Bacco G, Bianchi N. Analytical approach to determine the power limit of high-speed synchronous reluctance machines. In: 2017 IEEE International Electric Machines and Drives Conference (IEMDC). 2017. pp. 1-7

9

Adaptive Control for Estimating Insulation Resistance of High-Voltage Battery System in Electric Vehicles

Yi-Hsien Chiang and Wu-Yang Sean

Abstract

To ensure electrical safety and reliability in electric vehicles equipped with a high-voltage battery pack, an insulation monitoring circuit is indispensable to continuously monitor the insulation resistance during charging or driving. Existing methods such as injecting specific signals into the monitoring circuit and earth help to extract the resistance value from the voltage waveform. However, parasitic or stray capacitances in the monitoring circuit, which might introduce higher order dynamics into the waveform, are ignored. To avoid estimation error, the insulation resistance must be known in advance to carry out parameter calibration. In this chapter, one parasitic capacitance is applied in the circuit model and a new adaptive algorithm based on Lyapunov stability is employed to estimate the insulation resistance. This new online monitoring method and circuit are verified through simulation and experimentation, respectively. The results demonstrate that the proposed method can quickly react and track variations of insulation resistance on both positive and negative direct current (DC) lines.

Keywords: insulation resistance, adaptive control, electric vehicle, high-voltage battery system, reliability

1. Introduction

For improving efficiency, an increasing number of devices are operated at high-voltage levels to reduce losses in power transmission and power conversion. In the case of energy and power systems for electrical vehicles (EVs), a high-efficiency electrical device can be achieved using a high-voltage design. State-of-charge (SoC) and state-of-health (SoH) define the capability and reliability of a high-voltage battery, respectively. To determine these two parameters instantaneously, it is required to develop a simple, training-free, and easily implemented scheme.

Based on an equivalent-circuit model (ECM), the electrical performance of a battery can be formulated into a state-space representation. Besides, underdetermined model parameters can be arranged linearly so that an adaptive control approach can be applied [1]. However, electric shock may be harmful to passengers if insulation failure occurs. As insulation resistance in EVs varies with the operation environment and the reliability of the dielectric material [2], an online insulation-fault detection method is required to ensure that the insulation resistance stays within safe limits before startup or during the operation of high-voltage systems.

To ensure electrical safety in high insulation resistance, insulation inspection technologies have been widely applied in distribution networks [3–5] and EVs [6–13]. To analyze the dynamic insulation signals, the monitoring circuit is conventionally modeled by using two individual resistors connected with a DC high-voltage line to the earth ground. The basic concept used in estimating these two resistance values is to connect an additional circuit bridge composed of resistors and switches to obtain the differential current loops and corresponding voltages [8–11], but the resistance can only be measured if no current is conducted. As a result, these methods are barely suitable even for offline detection. A more practical approach for online detection is to continuously inject an excitation pulse signal into the negative terminal through a capacitor. The insulation resistance value can be acquired by analyzing the time-constant or the amplitude of the voltage waveform [12, 13]. However, parasitic or stray capacitance generally occurs in insulation loops, and causes considerable estimation error if they are ignored in the circuit model. In this chapter, we employ one typical capacitance in the monitoring circuit and propose a new adaptive algorithm for estimating both the resistance and capacitance values.

2. Topology of insulation monitoring circuit

To follow the process illustrated by Sottile and Tripathi [14], the high-voltage system we consider in this chapter could be an EV driving system that is powered by a high-voltage battery or a UPS that stores electricity to provide emergency power. As illustrated in **Figure 1**, a high-voltage system consists of a high-voltage battery pack, an inverter, a converter, an AC load, and an AC source. The DC sides of the inverter and converter are connected in parallel with the high-voltage battery pack. The AC side of the inverter is connected to a two-phase or three-phase load. In other cases, the AC side of the converter could be connected to a two-phase or three-phase source. Here, to consider an EV power system, we assume the AC load to be a traction motor driven by an inverter, with the converter acting as a charger that converts AC power from the grid or generator to DC power for charging the battery pack. In a UPS or a renewable energy system, however, both the AC load and source have the same grid network. Moreover, the inverter must be properly controlled to deliver power in phase with the grid power waveform.

The DC power line of the circuit connected to the high-voltage system is electrically isolated from the enclosure, i.e., the ground or chassis. Thus, we can determine the insulation status by measuring or estimating the resistance between the node on the positive-voltage line (+) or

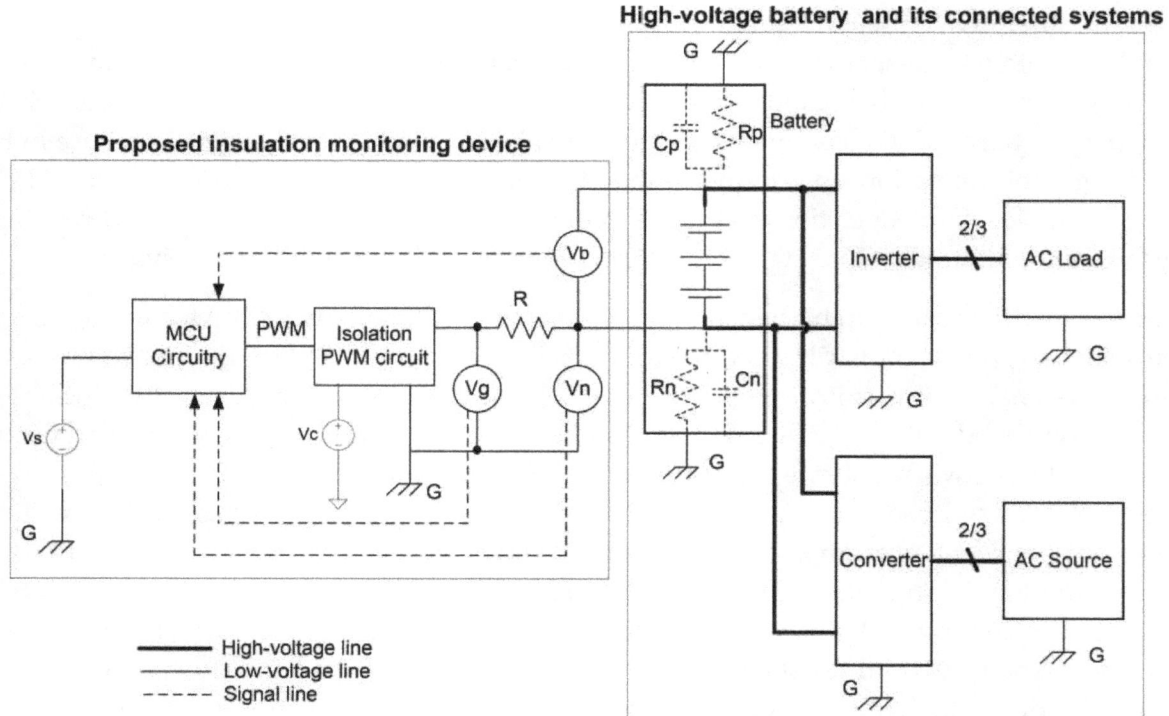

Figure 1. Proposed insulation monitoring circuit.

negative-voltage line (−) and the node with the equivalent electric potential to the ground. The electrical insulation in such a high-voltage system can be simply modeled by a resistor and a stray capacitor in a parallel connection, as shown in **Figure 1**, where R_p/R_n and C_p/C_n denote the positive/negative-line resistance and capacitance of the high-voltage system, respectively. We note that stray capacitance, which has been essentially ignored in previous studies, may yield considerable error in the RC circuit of the voltage waveform. In this work, we consider this higher-order dynamic response in our estimation model, and thus expect a more accurate result. To address this situation, our proposed insulation monitoring circuit, powered by the low-voltage battery V_s, has two outputs that are connected to the positive and negative nodes of the high-voltage system, as shown in **Figure 1**. In the monitoring circuit, we programmed an MCU (micro-control unit) to generate a PWM (pulse-width modulation) signal with a random duty-width sequence $d(t)$, which can be obtained by the PRBS (pseudo random binary sequence) method, which is widely used in systematic identification processes to fully excite the system dynamics. One example of a PWM signal is that shown in **Figure 2**, where $d(t)$ denotes the time-varying duty of the PWM, T is the time period, and $V_w(t)$ is the PWM output with ON and OFF voltage levels. The isolated PWM circuit generates random magnitude voltages in response to the random duty cycle sent from the MCU. Several topologies can achieve isolation and multiple-voltage generation on demand. For example, a photo-coupler provides electrical isolation and regulates the time for charging the output capacitor Cf, as illustrated in **Figure 3(a)**. We use the diode D_1 between the Rf and the external R to prevent any indirect connection to the ground via the Rf when the photo-coupler is turned ON. More specifically, we impose an R + Rf insulation resistance for the system when the coupler is

$$0 < d(t) \cdot T < T$$

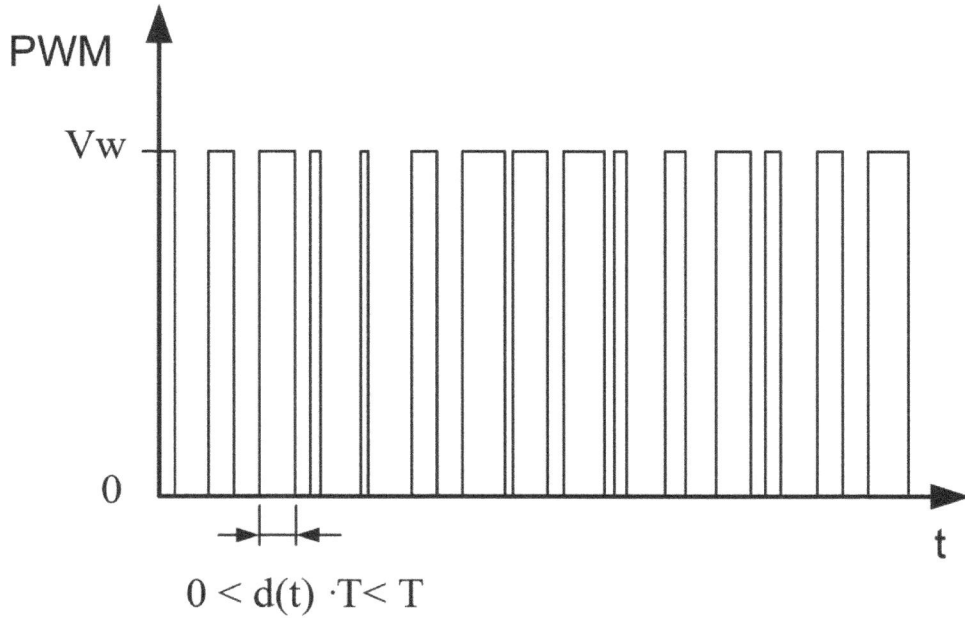

Figure 2. PWM signal generated using the PRBS method. (a) Photo-coupler circuit for electrical isolation and multiple-voltage generation. (b) General topology of the isolated DC/DC converter.

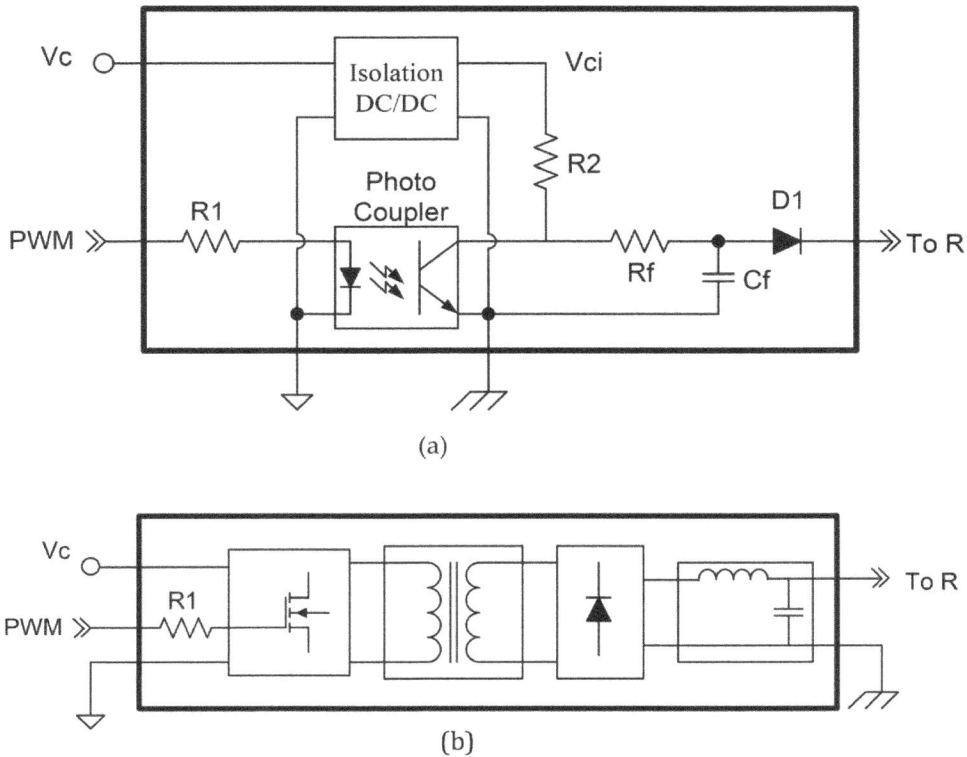

(a)

(b)

Figure 3. Several topologies for implementing the proposed isolation circuit.

turned ON, without diode D1 as a barrier. The other feasible topology is to isolate the DC/DC converter, such as a forward converter, a fly back converter, or a push–pull converter, as shown in **Figure 3(b)**. With this method, V_c is a DC source that can be generated from the external low-voltage source V_s. Then, the random voltage signal passes through a resistor R and is injected into a node on the negative line of the DC link. Three voltage measurements are required in this method, as shown in **Figure 1**. V_b is the voltage of the high-voltage system. V_n is the voltage between the high-voltage negative terminal and the ground, and V_g is the output voltage of the isolation circuit. The MCU continuously reads the instantaneously measured voltage for the online estimation of the insulation resistance R_p as well as R_n, using the derived adaptive control algorithm.

3. Equivalent circuit model

This high-voltage system connected to the insulation monitoring circuit can be modeled as an equivalent circuit, as illustrated in **Figure 4**, where V_b is the voltage of the high-voltage battery pack, V_a is the voltage of the two- or three-phase AC source or AC machine, the inverter/converter block is the power electronic circuit used to convert power between the AC and DC power stages, and V_g is the output voltage of the isolation circuit. R_p/R_n is the insulation resistor between the ground and the positive/negative terminal of the high-voltage battery pack. C_p and C_n are stray capacitors connected in parallel with R_p and R_n, respectively. Resistor R, which connects the positive terminal of the random voltage sources V_g and the negative terminal of the battery voltage V_b, forms a closed loop between the monitoring circuit and the high-voltage system. We note that the battery voltage V_b, the ground G, and the other insulation resistors and parallel capacitors form the other loop in the circuit.

Therefore, we estimate the insulation resistances R_n and R_p online by an algorithm that is based on the adaptive control law. According to Kirchhoff's circuit laws, the equivalent circuit shown in **Figure 4** can be described as follows:

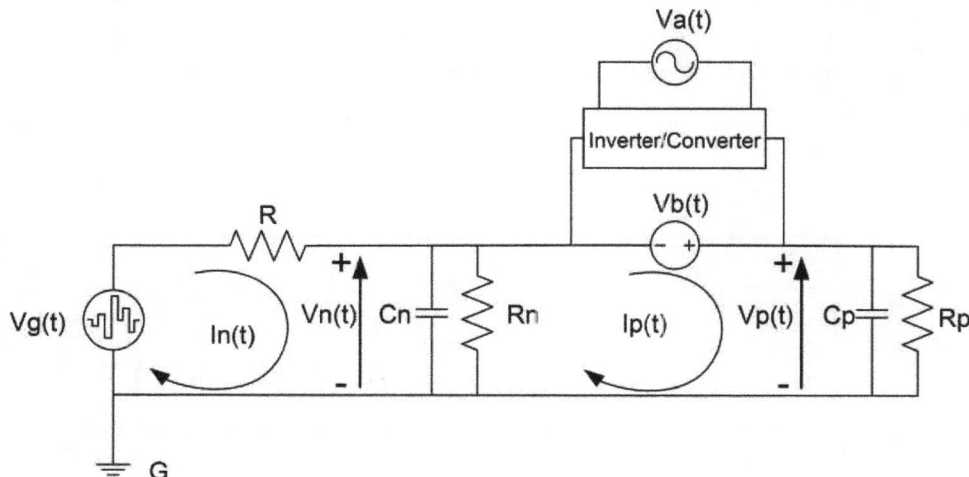

Figure 4. Equivalent circuit model for the insulation monitoring system.

$$I_P = C_P \dot{V}_P + \frac{V_P}{R_P} \tag{1}$$

$$I_N - I_P = C_N \dot{V}_N + \frac{V_N}{R_N} \tag{2}$$

Substituting Eq. (1) into Eq. (2), together with

$$V_P = V_B + V_N,$$

yields:

$$
\begin{aligned}
I_N &= C_N \dot{V}_N + \frac{V_N}{R_N} + I_P \\
&= C_N \dot{V}_N + \frac{V_N}{R_N} + C_P \dot{V}_P + \frac{V_P}{R_P} \\
&= C_N \dot{V}_N + \frac{V_N}{R_N} + \frac{V_P}{R_P} + C_P(\dot{V}_B + \dot{V}_N) \\
&= (C_N + C_P)\dot{V}_N + C_P \dot{V}_B + \frac{V_N}{R_N} + \frac{V_P}{R_P},
\end{aligned}
$$

which can be rewritten as follows:

$$
\begin{aligned}
\dot{V}_N &= -\frac{1}{R_N(C_N + C_P)}V_N - \frac{1}{R_P(C_N + C_P)}V_P + \frac{1}{C_N + C_P}I_N - \frac{C_P}{C_N + C_P}\dot{V}_B \\
&= -\left(\frac{1}{R_N} + \frac{1}{R} + \frac{1}{R_P}\right)\frac{1}{C_N + C_P}V_N + \frac{1}{R(C_N + C_P)}V_G - \frac{1}{R_P(C_N + C_P)}V_B \\
&\quad - \frac{C_P}{C_N + C_P}\dot{V}_B
\end{aligned}
\tag{3}
$$

Let us define the parametric vector as follows:

$$
\begin{aligned}
\theta^T &= \left[\frac{1}{C_N + C_P}\left(\frac{1}{R} + \frac{1}{R_N} + \frac{1}{R_P}\right) \quad \frac{1}{R(C_N + C_P)} \quad \frac{1}{R_P(C_N + C_P)} \quad \frac{C_P}{C_N + C_P}\right] \\
&= [\theta_1 \quad \theta_2 \quad \theta_3 \quad \theta_4]
\end{aligned}
\tag{4}
$$

and the variable vector as:

$$X^T = \begin{bmatrix} -V_N & V_G & -V_B & -\dot{V}_B \end{bmatrix} \tag{5}$$

such that the dynamics of the insulation monitoring system are formulated as follows:

$$\dot{V}_N = X^T\theta, \tag{6}$$

where the parametric vector includes all the resistance and capacitance values that must be known and the variable vector is composed of the variables that can be evaluated from all the measurements in the system, i.e., V_G, V_N, and V_P.

4. Proof of adaptive algorithm

If we suppose all the actual parameter values and the voltage V_N are unknown, we can write the dynamic model in an estimated formation as follows:

$$\dot{\widehat{V}}_N = \left[-\widehat{V}_N V_G - V_B - \dot{V}_B \right]\widehat{\theta} + u, \tag{7}$$

where $\widehat{Y}$ denotes the estimation of Y, and u is one part of the adaptation law that lets all the estimated values approach their true values, i.e., $\lim_{t\to\infty} \left| \widehat{V}_N(t) - V_N(t) \right| < \delta$ and $\lim_{t\to\infty} \left\| \widehat{\theta}(t) - \theta \right\| < \varepsilon$. We define the estimated error for V_N and the parametric vector as $e = V_N - \widehat{V}_N$ and $\widetilde{\theta} = \theta - \widehat{\theta}$, respectively. If we differentiate the estimated error, we have:

$$\dot{e} = \dot{V}_N - \dot{\widehat{V}}_N = \left[-\widehat{V}_N V_G - V_B - \dot{V}_B \right]\widetilde{\theta} - \theta_1 e - u. \tag{8}$$

Invoking the Lyapunov stability criteria shows that the positive-definite function:

$$S = e^2 + \widetilde{\theta}^T \Sigma \widetilde{\theta} \tag{9}$$

will approach zero for the negative semi-definite of its derivative; that is:

$$\dot{S} = \dot{e}e + \dot{\widetilde{\theta}}^T \Sigma \widetilde{\theta} = \left[-\widehat{V}_N V_G - V_B - \dot{V}_B \right]e\widetilde{\theta} - \theta_1 e^2 - ue - \dot{\widehat{\theta}}^T \Sigma \widetilde{\theta}$$

$$= -(\theta_1 + \lambda)e^2 + \left(\left[-\widehat{V}_N V_G - V_B - \dot{V}_B \right]e - \dot{\widehat{\theta}}^T \Sigma^{1/2} \right)\Sigma^{1/2}\widetilde{\theta} \tag{10}$$

$$= -(\theta_1 + \lambda)e^2 < 0,$$

provided that the adaptation law is as follows:

$$\dot{\widehat{\theta}} = \Sigma^{-1/2} \begin{bmatrix} -\widehat{V}_N e \\ V_G e \\ -V_B e \\ -\dot{V}_B e \end{bmatrix}, u = \lambda e^2, \text{ and } \lambda > 0, \tag{11}$$

where Σ could be a positive diagonal matrix for design simplicity. We can compute the insulation resistance as follows:

$$\widehat{R}_P = \frac{\widehat{\theta}_2}{\widehat{\theta}_3}R \text{ and } \widehat{R}_N = \frac{1}{\left(\frac{\widehat{\theta}_1}{\widehat{\theta}_2} - 1\right)\frac{1}{R} - \frac{1}{\widehat{R}_P}}. \tag{12}$$

Figure 5 shows a calculation flowchart for estimating the insulation resistance. A detailed description of the process is as follows:

i. Start the online estimation at time t_0. The initial values of the estimated insulation resistance can be updated with the latest value in memory for faster convergence.

ii. The voltage values are acquired from the measured V_g, V_n, and V_p values.

iii. The estimated voltage error is computed by Eq. (7) together with the updated parameters, where the initial value of the estimated V_n can be identical to the measurement.

iv. Based on the measured voltage data and estimated V_n, the adaptive algorithm given in Eq. (11) updates the parameter. The updated parameters are sent to the previous and following steps.

v. The insulation resistances are calculated by Eq. (12) based on the updated parameters. The minimum value is used to check whether it is under the predetermined threshold. If so, it is shown in the indicator.

vi. The waiting time required for the parameters' convergence is calculated and T_m is empirically set up to avoid misjudgment.

vii. Either the insulation resistance or an alert message is displayed, depending on whether it is below the mandatory threshold.

viii. For continuous online monitoring, once started, this flow is an infinite loop.

5. Simulation and experimental results

To verify the proposed algorithm, for simplicity, we assumed a scenario in which an electric vehicle is driven on the road such that the battery and AC line voltages are $V_b = 350$ V and $V_a = 0$ V, respectively. We set the initial insulation resistances at the positive and negative terminals to earth to be within the safety level, i.e., $R_p(t = 0\ \text{s}) = 600$ kΩ and $R_n(t = 0\ \text{s}) = 500$ kΩ, respectively. After 60 s, we degraded $R_n(t = 60\ \text{s})$ to 100 kΩ, and after 120 s, we did the same for $R_p(t = 120\ \text{s})$. To precisely characterize the electrical behavior of the equivalent circuit shown in **Figure 4**, we considered their parallel parasitic capacitances to be invariant at $C_p = 0.3$ uF and $C_n = 0.2$ uF. For the insulation resistance monitor, we selected the resistor R to be 20 kΩ, and we initially guessed the estimated values for R_p/R_n to be 350 kΩ/100 kΩ.

We constructed the circuit model and the estimation algorithm using Simulink software. The simulation estimation results for R_p and R_n are depicted in **Figures 6** and **7**, respectively. These figures show that the estimated R_p approaches the actual value within 20 s, but the estimated R_n converges to the actual value after 50 s. The relative error between actual and estimated values are both less than 1%. With respect to two degradation cases that sequentially occur on the negative and positive terminals, we found that either of the degradations would yield some fluctuation in the estimated value on the opposite side, particularly a case in which the degradation occurs on the positive terminal. As a consequence, it requires more time for convergence, i.e., 20 and 240 s for the degradations occurring on the negative and positive

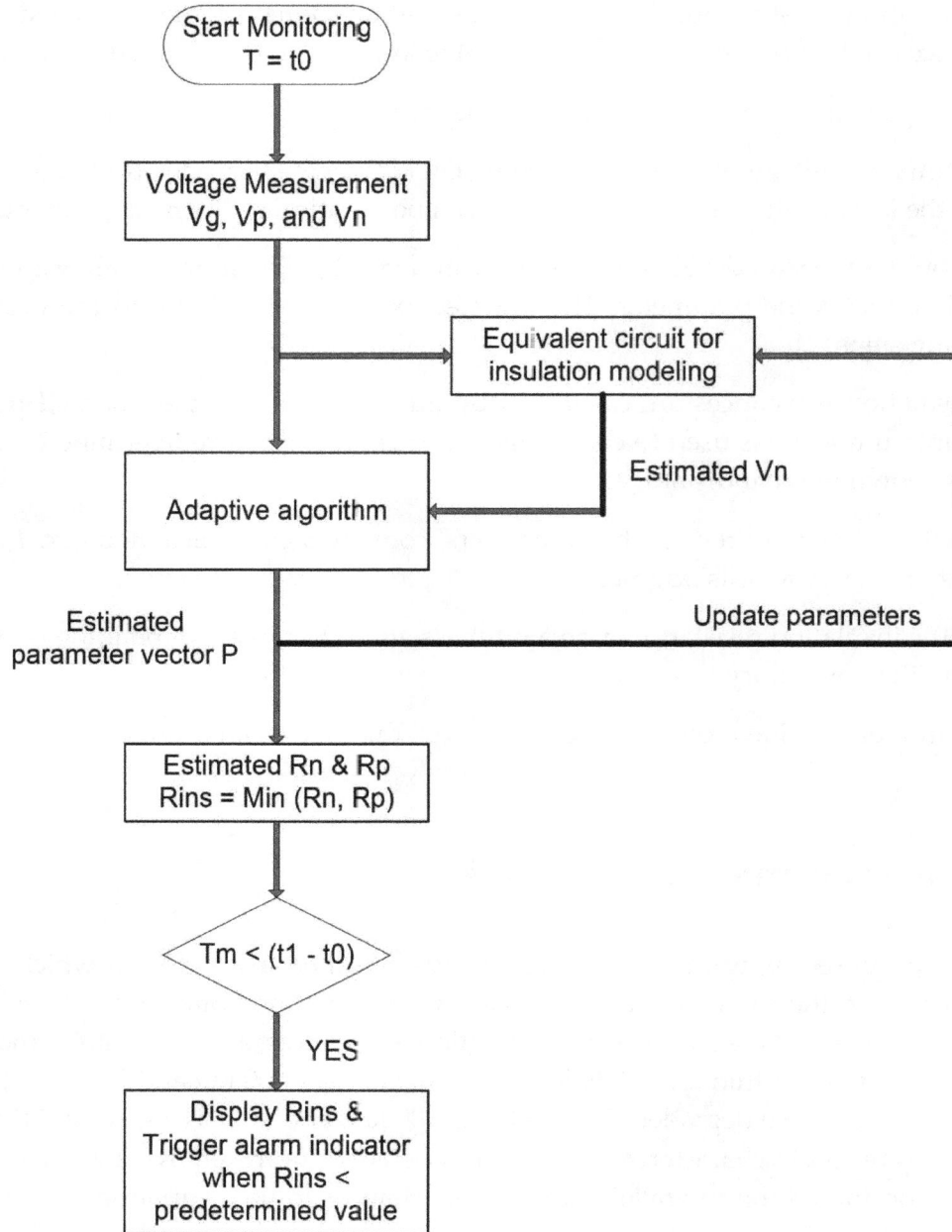

Figure 5. Flowchart for online parameter estimation.

sides, respectively. This is because the proposed circuit is directly connected to the negative terminal, which makes it more sensitive to voltage variations across the negative terminal and chassis ground. In other words, the high battery voltage V_b would attenuate the excitation signal coming from the negative side. Nevertheless, the simulation verifies that the proposed algorithm can estimate the actual insulation resistance and monitor its variation in the circuit model, while also considering the parasitic capacitance, as shown in **Figure 4**. To avoid false alarms due to ground fault detection when using this method, fault counting is necessary over a period of time.

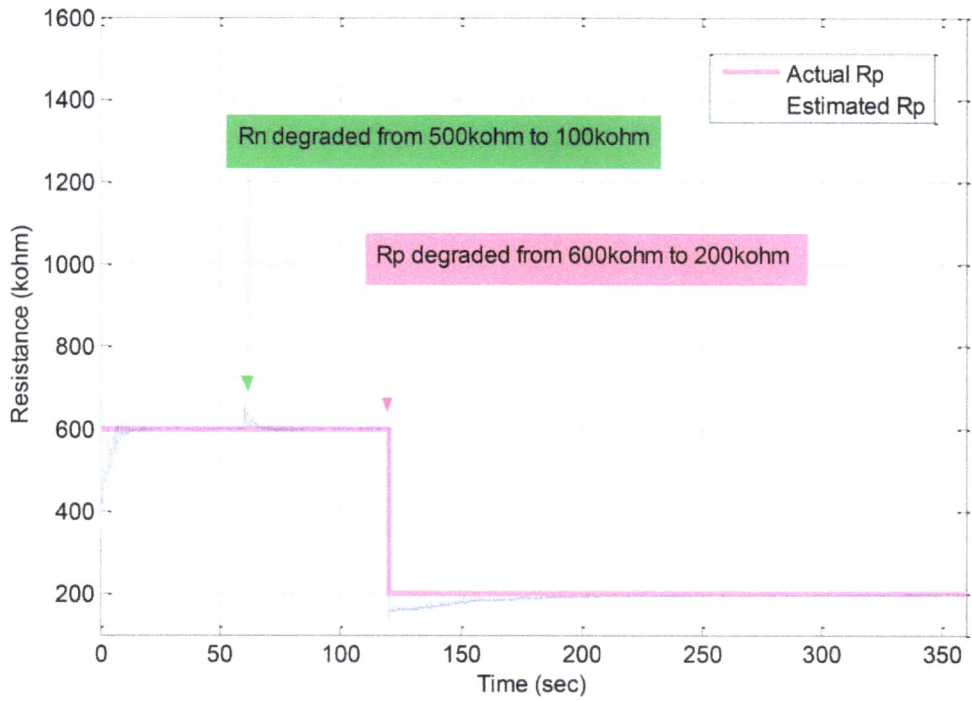

Figure 6. Estimation of R_p.

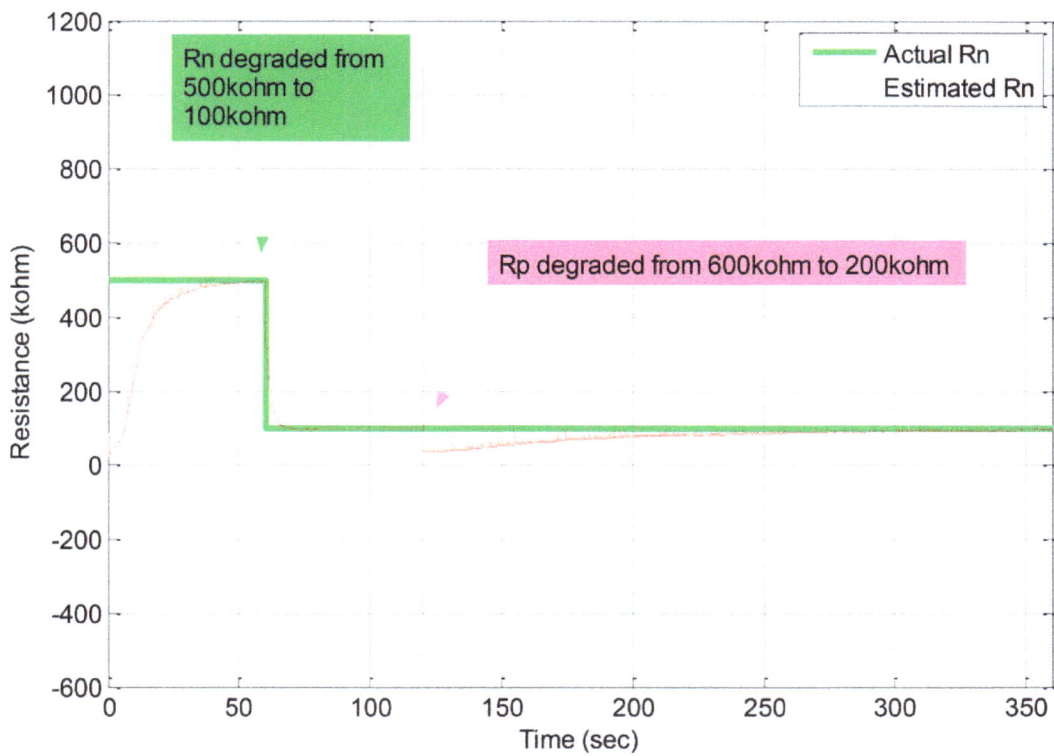

Figure 7. Estimation of R_n.

To simply validate the proposed algorithm in the laboratory, we connected a variable resistor to the proposed circuit to form a left-hand side loop of the circuit shown in **Figure 4**, in which the resistance, as represented by R_n, is the value to be estimated. Due to the simplicity of the single loop circuit, using Kirchhoff's current law, R_n can be evaluated in a straightforward manner, as follows:

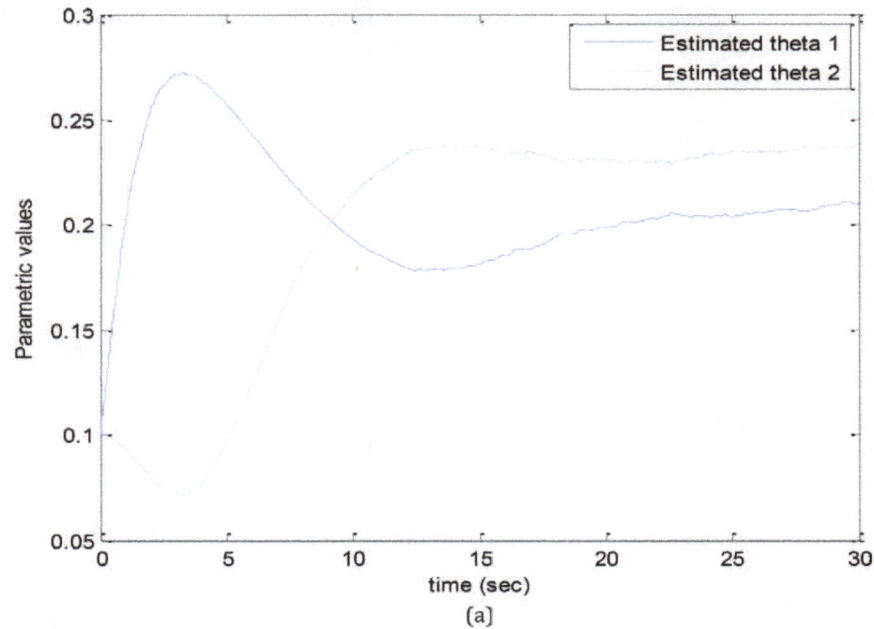

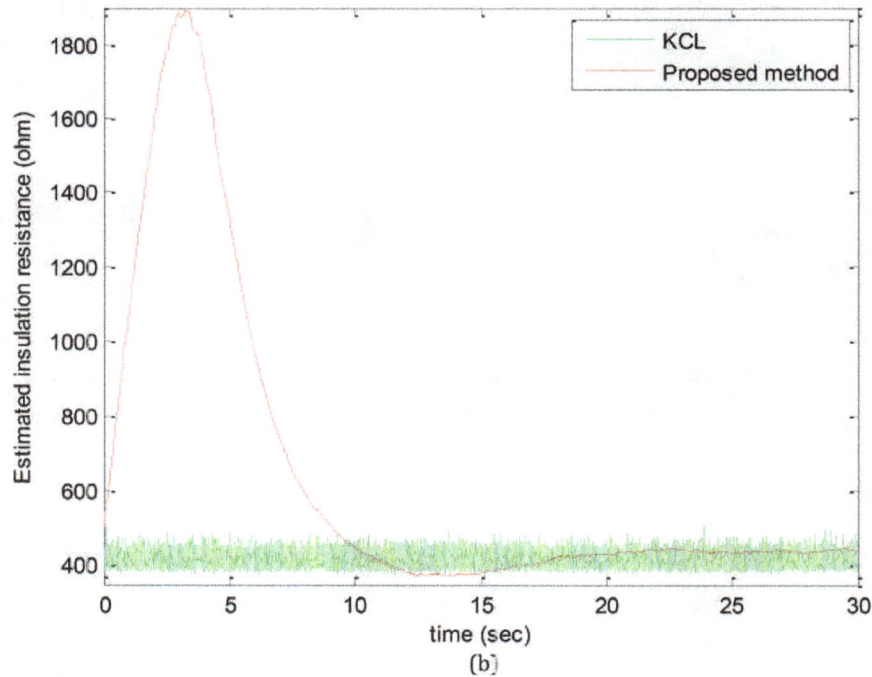

Figure 8. Experimental results. (a) Estimated parameters $\hat{\theta}_1$ and $\hat{\theta}_2$. (b) Estimated resistances of the two models.

$$R_n = \frac{R}{\frac{V_G}{V_N} - 1} \tag{13}$$

On the other hand, we modify the estimated model to yield:

$$R_n = \frac{\widehat{\theta}_1}{\widehat{\theta}_2} R. \tag{14}$$

The experimental results are shown in **Figure 8**. In **Figure 8(a)**, the two estimated parameters converge after 25 s. In **Figure 8(b)**, we depict the online estimated resistances based on the straight evaluation of Eq. (8) and the proposed method in Eq. (9). It is realized that the estimated value by using the straight evaluation varies roughly 10% between its maximum and minimum values. This may be due to either measurement noise or the dynamic uncertainty of the parasitic capacitance. However, the proposed method shows a steadier and more exact estimation after the convergence of the model parameters.

6. Conclusions

In this chapter, to improve existing techniques for enhancing the safety and reliability of high-voltage systems, we proposed a new insulation resistance online monitoring method for EV high-voltage DC lines, which takes into account the parasitic capacitance effect. The estimation scheme based on an adaptive control algorithm guarantees the asymptotical convergence of the parameters in the circuit model. Hence, as demonstrated in our simulation and experimental results, this method can steadily and accurately track the insulation resistance even when the parasitic capacitance is unknown. Due to the simplicity of the proposed algorithm and circuit, they can be easily implemented via electronic circuit design in real cases. According to the results, the estimated R_p and R_n converge to the actual value in 50 s. The relative error between actual and estimated values are both less than 1%. With respect to two degradation cases that sequentially occur on the negative and positive terminals, it requires more time for convergence, i.e., 20 and 240 s for the degradations occurring on the negative and positive sides, respectively.

Nomenclature

AC   alternating current

BMS   battery management system

DC   direct current

DC/DC   conversion of a DC source from one voltage level to another

ECM   equivalent-circuit model

EV   electric vehicle

MCU   micro-control unit

PRBS   pseudo random binary sequence

PWM   pulse-width modulation

SoC   state of charge

SoH   state of health

UPS   uninterrupted power supply

Author details

Yi-Hsien Chiang and Wu-Yang Sean*

*Address all correspondence to: wuyangsean@gmail.com

Department of Environmental Engineering, Chung Yuan Christian University, Taiwan

References

[1] Stimper K. Physical fundamentals of insulation design for low-voltage equipment. IEEE Transactions on Electrical Insulation. 1990;**25**(6):1097-1103

[2] Baldwin T, Renovich F, Saunders LF. Direction ground-fault indicator for high-resistance grounded systems. IEEE Transactions on Industry Applications. 2003;**39**(2):325-332

[3] Huang H-H, Quan C, Huang J. Development of distributed DC grounding detecting system based on differential current detecting method. Journal of Electronic Measurement and Instrument. 2009;**23**(11):8

[4] Li L-W, Liu X-F, Liu B. Distributed on-line grounding monitoring system for DC system based on field bus. Electric Power Automation Equipment. 2006;**26**(12):55-58

[5] Guo H-Y, Jiang J-C, Wen J-P, Wang J-Y. New method of insulation detection for electrical vehicle. Journal of Electronic Measurement and Instrument. 2011;**25**(3):253-257

[6] Kota O, Balasubramanian G. High voltage safety concepts for power electronic units. SAE Technical Paper; 2013

[7] Li J-X, Fan Y-Q, Jiang J-C, Chen H. An approach to on-line monitoring on insulation resistance in electric vehicle. Automotive Engineering. 2006;**28**(10):884-887

[8] Li L, Jiang J-C. Research on battery insulation detection for electric vehicle. Electronic Measurement Technology. 2009;**32**(2):76-78

[9] Pan L, Jiang J, Li J. Development of intelligent passive grounding detection device for electric vehicle. Electrical Drive Automation. 2003;**25**(4):47-48

[10] Wu Z-J, Wang L-F. A novel insulation resistance monitoring device for hybrid electric vehicle. In: IEEE Conference on Vehicle Power and Propulsion (VPPC); 2008

[11] Morimoto N. Ground-fault resistance measurement circuit and ground-fault detection circuit. US Patent No. 7560935B2

[12] Onnerud P, Linna JR, Warner J, Souza C. Safety and performance optimized controls for large scale electric vehicle battery systems. US Patent No. 20110049977A1

[13] Chiang Y-H, Sean W-Y, Huang C-Y, Hsieh L-HC. Adaptive control for estimating insulation resistance of high-voltage battery system in electric vehicles. Environmental Progress & Sustainable Energy. 2017;**36**(6):1882-1887

[14] Sottile TN, Tripathi AK. Best practices for implementing high-resistance grounding in mine power systems. IEEE Transactions on Industry Applications. 2015;**51**(6):5254-5260. DOI: 10.1109/TIA.2015.2420632

10

Worked Example of X-by-Wire Technology in Electric Vehicle: Braking and Steering

Ameer Sider and Mohd Khair Hassan

Abstract

The chapter emphasizes on the worked example of braking system and steering system for electric vehicle. The x-by-wire technology is investigated and validated comprehensively. Brake-by-wire is considered a new brake technology that uses electronic devices and control system instead of conventional brake components to carry out braking function based on wire-transmitted information. However, the physical parameters associated with braking function cause nonlinear characteristics and variations in the braking dynamics, which eventually degrade stability and performance of the system. Therefore, this study presents the design of fuzzy-PID controller for brake-by-wire (BBW) to overcome these undesired effects and also to derive optimal brake force that assists to perform braking operation under distinct road conditions and distinct road types. Electric power-assisted steering (EPAS) system is a new power steering technology for vehicles especially for electric vehicles (EV). It has been applied to displace conventional hydraulic power-assisted steering (HPAS) system due to space efficiency, environmental compatibility, and engine performance. An EPAS system is a driver-assisting feedback system designed to boost the driver input torque to a desired output torque causing the steering action to be undertaken at much lower steering efforts.

Keywords: x-by-wire, brake-by-wire, steer-by-wire

1. Introduction

In recent times, potential energy, environment, and economic interests have stimulated motorized industry to develop and enhance efficient, clean, and sustainable vehicle, particularly, for city transportation. This new invention should not contingent on oil as a sole of energy source. Additionally, reducing engine size, replacing mechanical components by electrical devices, transferring request information electronically instead of mechanically, and designing integrated control systems are considered other targets that automotive manufactures are aiming to attain in producing new means of transportation. At that juncture, the automotive industry introduced electrical vehicle (EV), which is driven by alternative energy sources that provide magnificent means for efficient, clean, and environmentally urban transportation.

The trend technology toward electronic components and circuits coming from their technical merits not only reduces the weight of vehicles but also has the potential for a large number of integrated functions and features. Some of these new electronically operated systems are taken place under the concept of x-by-wire, which involves brake-by-wire, throttle-by-wire, and steer-by-wire. These electrical vehicle subsystems yet still undergo considerable challenging issues that need intensive study and investigation in order to find out appropriate design and powerful operated system.

This chapter presents x-by-wire technology implementation in electric vehicle. BBW is a new brake technology in which mechanical and hydraulic components of traditional brake systems are replaced by electric circuits and devices to carry out the function of braking in a vehicle by wire-transmitted information. The advantages of electronic devices such as reducing vehicle weight and increasing brake performance are considered the main purpose trends of the automotive industry toward this new brake technology. Another application known as n EPAS system is a driver-assisted feedback system designed to boost the driver input torque to a desired output torque causing the steering action to be undertaken at much lower steering efforts. Particle swarm optimization (PSO) algorithm is implemented as tuning mechanism for fractional-order PID (FOPID) controller. The aim of this controller is to track the assist current generated by lookup table. The results show the performance and efficiency of using PSO algorithm for FOPID tuning.

The motivation of this study is to enhance the safety aspects for the vehicle while attaining any desired speed. To achieve that, an optimal brake force at different road types and conditions and for different brake commands must be obtained within a reasonable time and without vehicle sliding.

2. BBW design and principle of work

The proposed design of BBW used in this study is schematically illustrated in **Figure 1**, which includes one wheel of vehicle model as seen inside the dotted box. According to the figure, the suggested principle of work of BBW is adopted which is demonstrated as follows:

Primarily, reducing (or halting) vehicle speed comes as a result of pressing down on the brake pedal by the driver. The braking pedal of BBW is usually equipped with several electronic sensors that provide redundant information about braking request. Thus, when a brake force applies to the brake pedal, three possible sensors are usually utilized to measure required braking force: (1) pedal displacement sensor (measures pedal displacement as a result of applying force on the pedal) [3], (2) force sensor (measures applied force on brake pedal), and (3) pressure sensor (measures applied pressure to brake pedal) [5]. In addition to that, the brake pedal of BBW may not necessarily be as the general brake device, rather than it could be a hand-adjacent device placed at the steering wheel that enables driver to apply brakes with

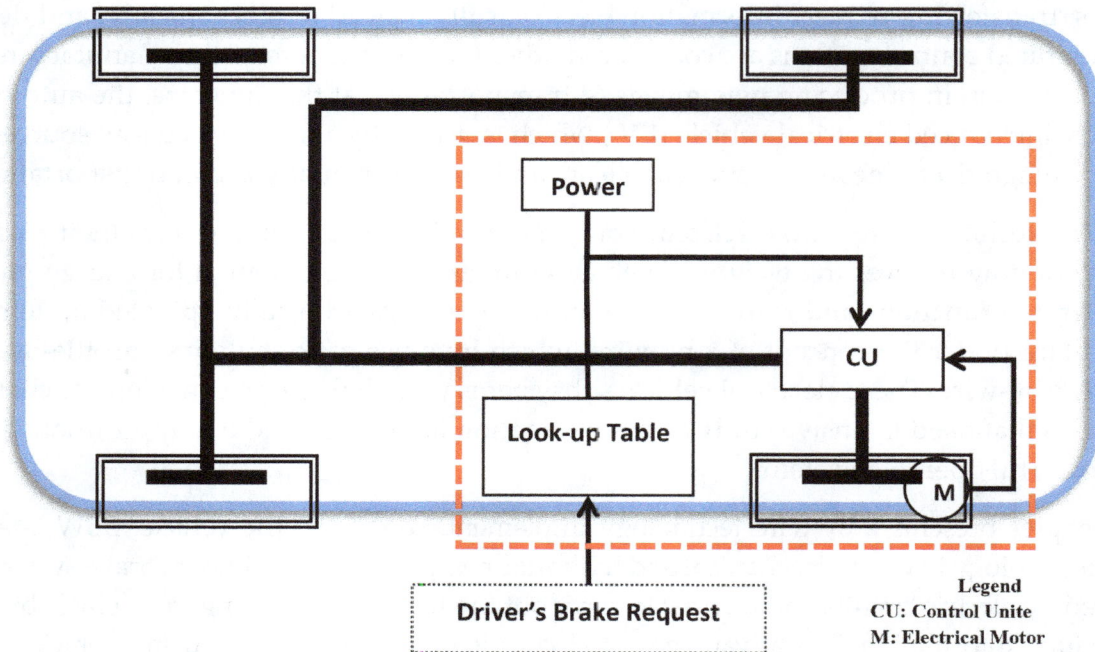

Figure 1. Proposed BBW architecture for one-wheel brake model.

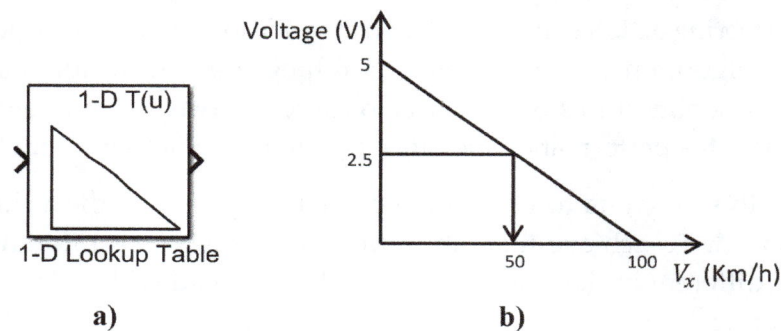

Figure 2. Lookup table of the input brake force. (a) Tool box. (b) Vehicle speed-voltage source.

hand movement as suggested in [5]. However, since the focus of this study is to design control strategy for braking action, it is assumed that braking request is already measured and available in the form of voltage source as adopted by Mingfei's design [1]. Therefore, the chosen voltage source of this study exists within a range of 0–5 V in which 0 V relates to released brake pedal and 5 V relates to fully pressed pedal. This voltage range is formulated in such a way that brake pedal input, which is in the form of voltage source, matches the desired vehicle speed by using 1D lookup table as shown in **Figure 2**. This lookup table enables a range of inputs that correspond to [0–5] Volts which in turn this range corresponds to the vehicle speed range [0–100] Km/h; for example, if the input brake pedal corresponds to 2.5 V, the relative required vehicle speed will be 50 Km/h as illustrated in **Figure 2(b)**. Nonetheless, the relative values of vehicle speed are changeable according to initial vehicle speed (vehicle speed before braking action), whereas voltage range remains constant all the time and has the capability to correspond to any given vehicle speed by updating vehicle initial speed. For instance, if the initial vehicle speed is set to 50 Km/h, the voltage range [0–5] V will correspond to the vehicle speed [0–50] Km/h as explained in **Figure 2(b)**.

Upon determining the required brake request, the braking command is then sent to the control unit (CU) via wires as shown in **Figure 1**. The CU located at the wheel after that determines exactly the control signal that must be transmitted to the brake actuator unite in order to slow down or stop the vehicle. Nevertheless, the control signal of the CU is considered the input for the electrical actuator (permanent magnetic DC motor) where this signal takes the form of the desired braking torque. Consequently, electronic actuator of the brake unit operates based on the desired braking torque which in turn decreases (or stops) vehicle speed according to the desired speed.

The control unit, however, is updated through feedback control strategies where wheel speed is considered the input to the feedback control system according to applied control strategy. Moreover, the interaction between brake pedal, control unit, electronic actuator, and wheel as well as vehicle speed is completely accomplished by wires. In view of that, vehicle brake system is designed and structured.

3. Control system design

The suggested control brake system employs fuzzy-PID controller to obtain the desired vehicle speed based on tuning of traditional PID controller. The applied control algorithm must be able to function to any required vehicle speed that is determined by the driver. The proposed control system design applied to handle this task is schematically illustrated in **Figure 3**.

As depicted in **Figure 3**, the system input (brake pedal force) is determined by the driver in the form of voltage signal ranging from 0 V (refers to release pedal) to 5 V (refers to fully pressed pedal). Upon determining the required vehicle speed by lookup table, the speed signal is then sent to the control unit which is based on either PID or fuzzy-PID controllers. The implemented control algorithm then determines the desired voltage source that must be transmitted to electrical actuator in order to generate required braking torque. The wheel speed after that is decreased by applying brake torque causing modification in overall system dynamics which in turn leads to vehicle speed deceleration.

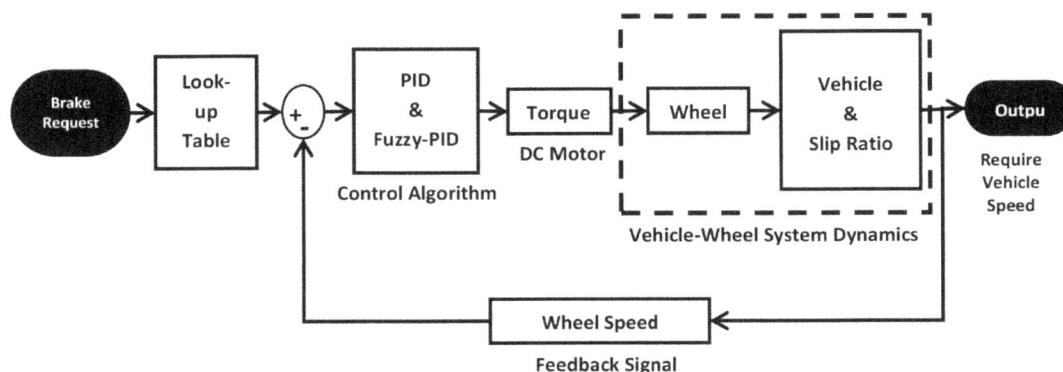

Figure 3. BBW control system design.

The error signal transmitted into control algorithm, however, is determined by the difference between input signal (desired vehicle speed) and feedback signal (wheel speed) which is given by the following relationship:

Error = required vehicle speed (input signal)–measured wheel (feedback signal)   (1)

The control strategy used to deliver the desired vehicle speed is based on maintaining peak slip ratio within the maximum adhesion characteristic range [0.02–0.35]. Locating peak slip ratio within the maximum friction characteristic initiated from applying ideal and accurate brake torque is capable of deriving proper and acceptable vehicle-wheel speed relationship.

The control objective of both controllers is to decrease vehicle velocity to the desired vehicle speed (5 Km/h) while maintaining slip ratio within its maximum range [0.02, 0.35]. Besides, the control algorithms are designed to operate braking action on dry asphalt road type, whereas other road types and conditions (such as wet asphalt, wet and dry cobblestone, and concrete) are applied to examine and investigate whether the controllers can handle characteristic variations of the system or not.

a) PID controller design

A cascade-form PID controller is designed based on manual tuning method, where the three terms of PID controller (proportional, integral, and derivative) are employed. Accordingly, the overall controller output is considered the sum of the contributions of the individual PID terms which is further expressed in Eq. (1), where $u(t)$ is the PID control signal, $e(t)$ is the error signal, and K_p, K_i, K_d are the proportional gain, integral gain, and derivative gain, respectively.

$$u(t) = K_p e(t) + K_i \int_0^t e(\tau)d\tau + K_d \frac{d}{dt} e(t)$$ (2)

b) Fuzzy-PID controller design

Although PID manual tuning method provides stable output response, PID controller does not achieve the desired control specifications since the dynamics of the system has nonlinear and variant parameters which in turn degrade system performance. Therefore, fuzzy logic controller has been introduced to PID controller in order to improve the response as well as to enhance system performance based on fuzzy-PID tuning. In fact, fuzzy-PID controller is considered as a link between traditional control which has well-established theory and intelligent control that conquers traditional control problems like nonlinearity.

Fuzzy-PID scheme, in addition, can employ different structures and forms based on the input to the fuzzy controller on the one hand and on the arrangement of PID parameters and their locations with respect to fuzzy controller on the other hand. Nonetheless, these different structures are possible in the context of knowledge description and explanation, whereas they should be examined with respect to their functional behavior. The proposed structure of this study is schematically illustrated in **Figure 4**, which generates incremental and absolute fuzzy-PID signal based on direct action to tune PID parameters through fuzzy inference.

As shown in **Figure 4**, the error and rate change of error are considered as the time-varying inputs to the fuzzy logic controller (linguistic inputs), whereas tuned (proportional, integral,

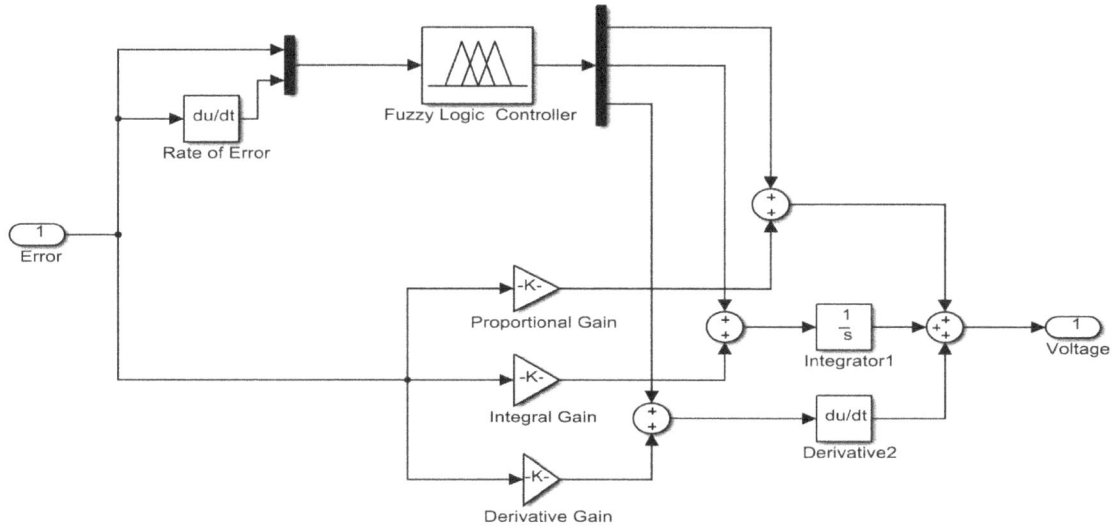

Figure 4. Fuzzy-PID controller (MATLAB Simulink scheme).

and derivative) gains are the output of the controller (linguistic outputs). Regarding linguistic inputs, there are other choices (such as integral of error) that could also be used as input variables, yet the selection variables make good intuitive sense, particularly as the input error is naturally engaged in the control problem of regulating process output around specific set point. The controller input variable however must have proper information available to provide good decision to derive vehicle speed into the desired speed to achieve high-performance operation based on fuzzy-PID tuning. On the other hand, the linguistic output variables are expressed as tuned (proportional, integral, and derivative) gains where the output values of these tuned gains are implemented to tune the conventional PID controller as shown in **Figure 4**.

The adopted linguistic values and their corresponding abbreviations in conjunction with their linguistic variables are summarized in **Table 1**.

This table provides a language to express the control decision-making process in the context of established input–output framework. For example, the statement "error is negative" can be referred to the situation where the vehicle speed curve exists above the desired speed and needs more braking force. In contrast, the statement "error is positive" can be referred to the

	Linguistic variables	Linguistic values	Linguistic value abbreviation
Input	Error	Positive, zero, and negative	P, Z, and N, respectively
	Rate change of error		
Output	Proportional gain	Positive, zero, and negative	P, Z, and N, respectively
	Integral gain	Zero, positive small, and positive large	Z, PS, and PL, respectively
	Derivative gain	Zero, positive small, and positive large	Z, PS, and PL, respectively

Table 1. Linguistic variables alongside their linguistic values and abbreviations.

	Error			
Change in error		N	Z	P
	N	N	N	Z
	Z	N	Z	P
	P	Z	P	P

Table 2. Fuzzy rule base for proportional gain.

	Error			
Change in error		N	Z	P
	N	Z	Z	PS
	Z	Z	PS	PL
	P	PS	PL	PL

Table 3. Fuzzy rule base for integral and derivative gains.

situation where the vehicle speed curve exists below the required speed curve and needs to decrease applying torque to obtain the desired vehicle speed.

Upon determining linguistic quantification, the rule base of the control system is set to capture expert's knowledge on how to tune the system and describe applied control strategy. Since there are two input variables and three output variables, the possible rules can at most reach to 3^2 (9) rules. These rules are listed in a tabular representation form as shown in **Tables 2** and **3**.

The meaning of the above linguistic description is quantified via membership function, whereas triangular shape is considered in this study for all inputs as well as all outputs for its simplicity, linear grade distribution, and fairly limited availability of the relevant information about the linguistic terms. In due course, the selected membership functions and their associated universe of discourse as well as linguistic values of this study are revealed in **Figure 5**. The designed membership functions are overlapped, and the height of the intersection of each two successive fuzzy sets is ½.

Since a clear picture on the linguistic variables, rule base, and membership functions have been explained, we move to the important issue of how the exact fuzzy controller works. In doing so, the first component of fuzzy controller is fuzzification process which is the act of acquiring the value of the input variable and defining numeric magnitudes for the membership function that are set for that variable. After that, the inference mechanism takes the action through two steps:

1. Matching the premise associated with all the rules to the controller inputs to determine which rules apply to the current condition. In other words, each rule in the rule base has different premise membership functions on the one hand and function of error and change

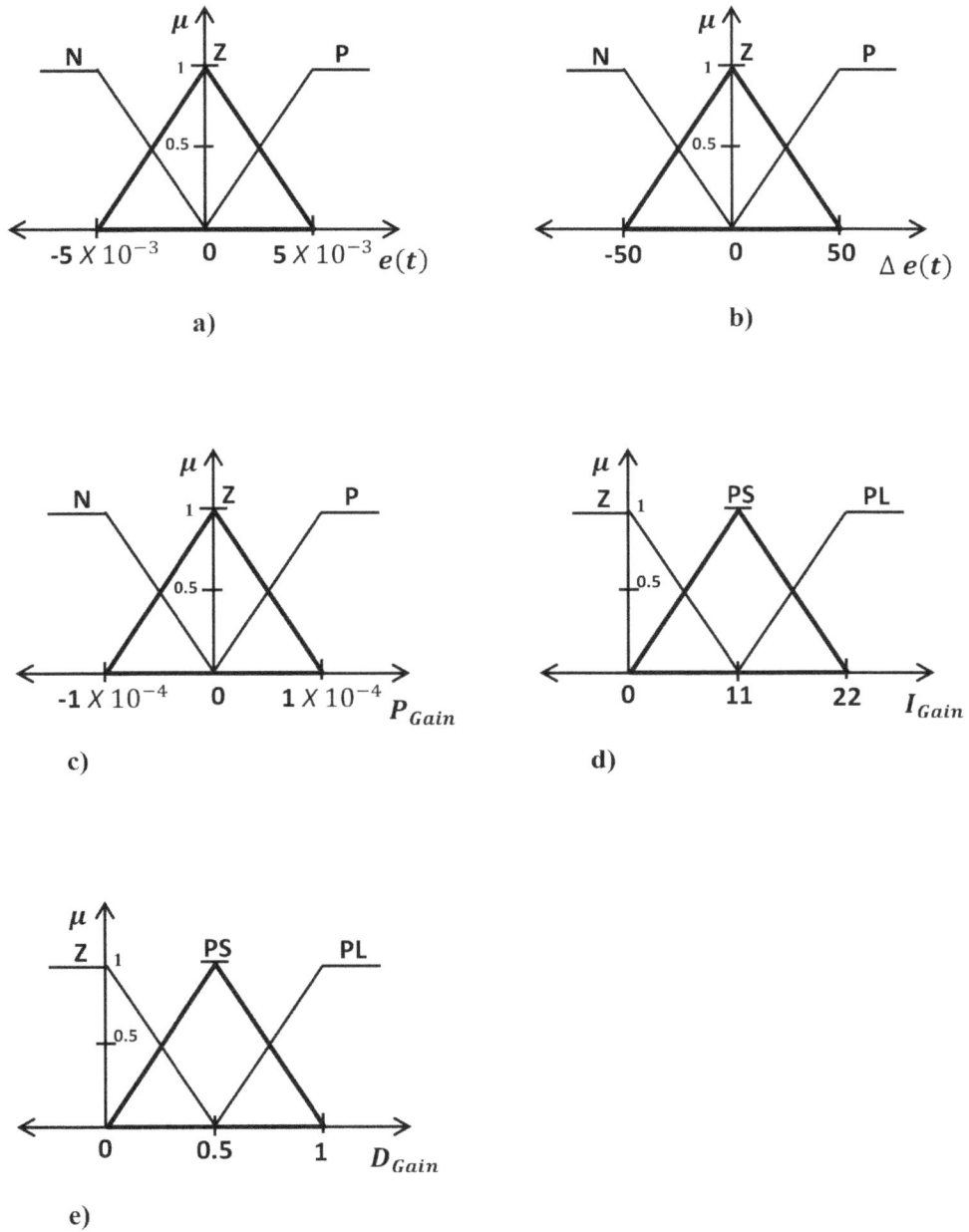

Figure 5. Membership functions and their corresponding values. (a) Membership functions and their values for error input $e(t)$. (b) Membership functions and their values for change of error $\Delta e(t)$. (c) Membership functions and their values for proportional gain. (d) Membership functions and their values for integral gain. (e) Membership functions and their values for derivative gain.

in error on the other hand; therefore, the quantification of the certainty that each rule base applies to the current condition can be obtained upon providing specific values for the error and change in error.

2. Determining the conclusion (what the control action to take) that should be applied by using selected rules to relate to the current situation. This conclusion is classified with a fuzzy set that signifies the certainty that the input to the plant should undertake various values.

Therefore, as long as the input to the inference process (set of rules) is on, its corresponding output operates which is in the form of implied fuzzy sets. However, these implied fuzzy sets are then converted to crisp values (numeric values) by combining their effects to give the most certain controller outputs. The defuzzification process is obtained by bisector method which divides the area by a vertical line into two equal subregion areas. In addition, the mean of maximum and the largest of maximum are also applied to the system for the purpose of validity in which both of them provide close output result.

4. Simulation result and analysis

The vehicle model and controller algorithms are examined in MATLAB software. For the results of investigation and analysis, the initial vehicle-wheel speeds are set to 100 Km/h, whereas the desired vehicle speed is set to 5 Km/h. The reason of choosing 5 Km/h as the desired vehicle speed instead of zero Km/h is because slip ratio magnitude goes to infinity as vehicle speed approaches zero which in turn leads to inappropriate output behavior. On the other hand, selecting the desired low speed helps to examine maximum slip ratio that controls algorithm derives; hence, the ability to evaluate control performance and output response of the system will be more effective and visible.

The output responses of fuzzy-PID controller for dry asphalt road type are presented in **Figure 6**, whereas **Figure 6(a)** demonstrates the output responses of vehicle-wheel, and **Figure 6(b)** shows the output responses of slip ratio. Yet, the output responses of traditional PID controller are imposed in the same figure (**Figure 6**) to illustrate the comparison between traditional PID controller and fuzzy-PID controller.

As shown in **Figure 6(a)**, both controllers could derive stable output response smoothly. However, the output performance of the fuzzy-PID controller is much better than conventional PID controller since PID controller derives large steady-state error on the one hand and takes long time (approximately 15 seconds) to approach the desired vehicle speed (5 Km/h) on the other hand. In contrast, fuzzy-PID controller overcomes these problems being provided better output performance with zero steady-state error. As a result, fuzzy-PID controller could obtain the required vehicle speed within approximately 9 seconds which in turn assists to reduce stopping vehicle time 60% as compared to PID controller and more importantly the ability of fuzzy-PID controller to eliminate steady-state error to zero. Therefore, fuzzy-PID controller shows superior and outstanding controller.

On the other side, the output response of slip ratio associated with vehicle-wheel speed as shown in **Figure 6(b)** reveals smooth output response particularly before attaining the desired output speed. As depicted from the figure, the maximum slip ratio is the same for both controllers which approximately equals to 0.027. Though the maximum slip ratio magnitude seems a small value, the main cause for vehicle-wheel deceleration is considered since friction force between road surface and wheel surface principally depends on the slip ratio magnitude even though if the slip ratio possesses very small magnitude that may reach to mili-slip ratio.

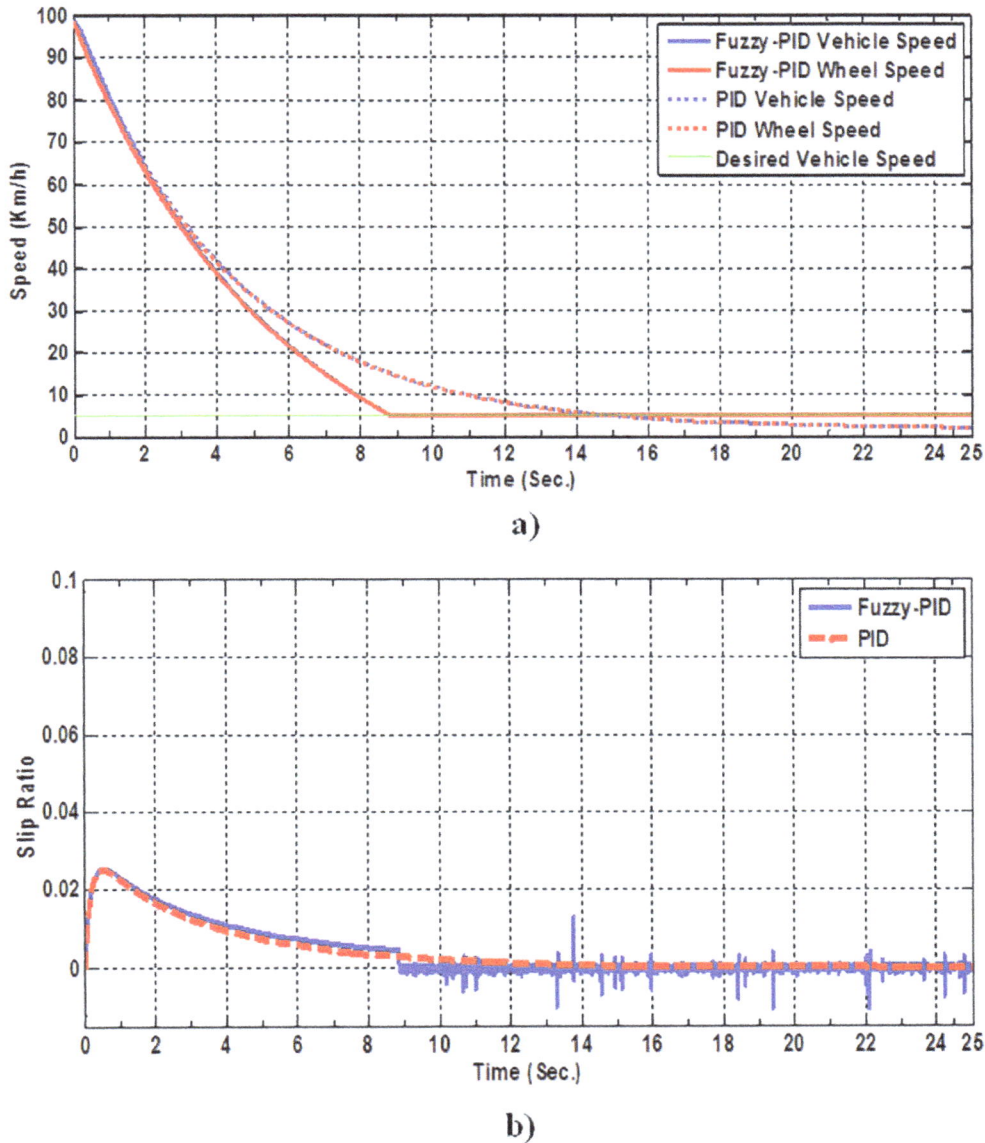

Figure 6. Output responses of fuzzy-PID controller for dry asphalt road type. (a) vehicle-wheel speed. (b) Slip ratio.

This fact is clearly observed from **Figure 6(b)**, especially within the time interval [2, 4] seconds, where the slip ratio of fuzzy-PID controller output response (blue line) has larger magnitude than PID output response (red line) by mili-values. This slight divergence that fuzzy-PID created, by trivial increases in slip ratio magnitude, leads to dramatically improve and enhance output response by decreasing vehicle stopping time 60% as compared to conventional PID controller.

As shown in the table, the slip ratio increases as the adhesion characteristic decreases. For instance, the maximum adhesion characteristic of dry asphalt is about 1.18 which is considered a large value; hence, its magnitude-derived slip ratio is small (0.027 for PID controller and 0.028 for fuzzy-PID controller). In contrast, the wet cobblestone adhesion characteristic has a small value (0.34), and therefore its derived slip ratio has a large value (0.26 for PID controller and 0.33 for fuzzy-PID controller).

Road-type controller-type	Max. adhesion char.	PID	Fuzzy-PID
Asphalt (dry)	1.18	0.027	0.028
Asphalt (wet)	0.8	0.035	0.035
Concrete	1.1	0.028	0.029
Cobblestone (dry)	1	0.085	0.087
Cobblestone (wet)	0.34	0.26	0.33

Table 4. Maximum derived slip ratios for PID and fuzzy-PID controllers.

The other significant notice that can be observed from **Table 4** is the maximum slip ratio of fuzzy-PID controller which is slightly larger than the one derived by PID controller, meanwhile fuzzy-PID performance is considered superior and much better than PID performance as demonstrated above. Accordingly, the slip ratio magnitude is considered extremely important in braking operation even though if it possesses very small value since braking process depends on the road-wheel surfaces. Nonetheless, a certain magnitude range of slip ratio is permissible where if its magnitude exceeds that range, the operating system may undergo unwanted behavior (wheel locks up or losses the control) particularly if it goes to a large value (more than 0.5).

It is also concluded that the mathematical derivation and its investigation of the brake system model are accurate and valid particularly because the examination and exploration of the output results are entirely identical to the analysis and investigation of each system's variable as demonstrated in Section 3. Besides, the suggested feedback control signal which is based on wheel speed was able to deliver detailed and thorough information about the status of braking system which assists to update system's variable effectively.

5. EPAS system

EPAS presents the continuing future of power-assisted steering technology for passenger vehicles and has already been started to appear in high-volume, lead-vehicle applications; more flexible than traditional hydraulic power-assisted steering (HPAS) system, the fact of EPAS is to supply steering assistance to the driver utilizing an electrically controlled electric motor. EPAS is a classic exemplary case of a smart actuator operating under feedback control. It can provide necessary assist torque in different car speeds and different driver torques [6]. It has been reported in [6] that among electric power-assisted steering (EPAS) system available for passenger cars, EPAS systems provide the best fuel consumption [7–9]. The plot shown in **Figure 7** indicates that EPAS systems have the lowest fuel consumption in comparison to hydraulic power-assisted steering (HPAS) system with savings in excess of 3.0% in average and up to 3.5% in city driving [6].

According to the steering torque, automobile speed as well as road conditions, the system can provide the real-time assistant torque through assist motor to help driver steering and make steering easier and gentle, which guarantees that the driver has the best steering feel in the variety of operating conditions. At present, the design for the assist motor control have mainly two methods: the first one is motor current loop control based on classical control theory and the other one is

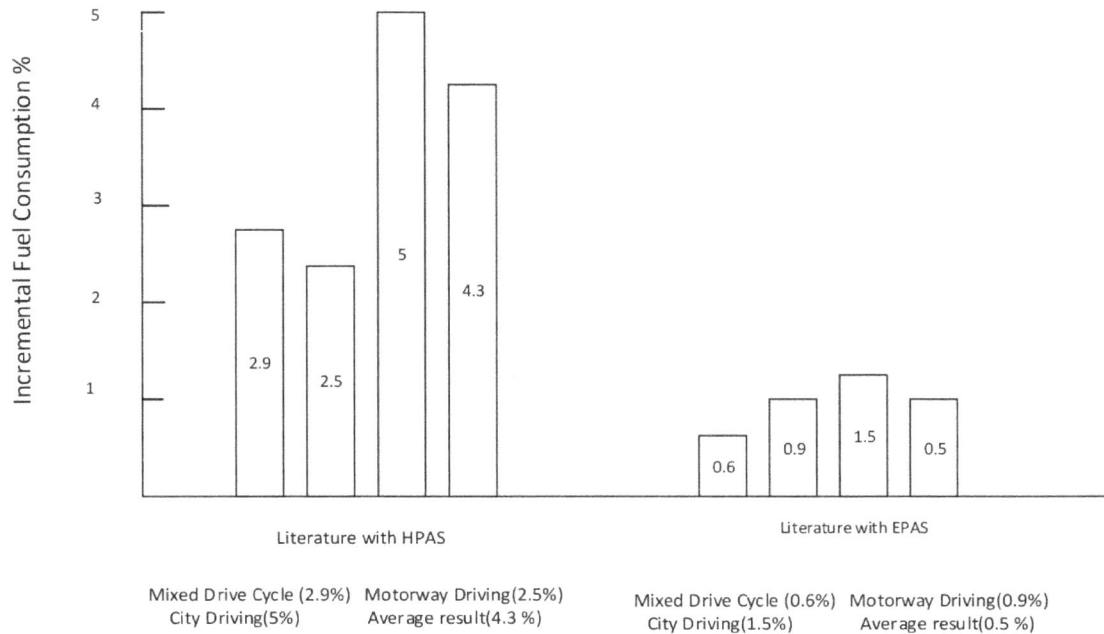

Figure 7. Typical EPAS fuel consumption saving.

the state-space model H_∞ control or sliding mode control based on modern control theory [10]. Literature [11] using the motor current tracking control based on conventional PID achieved good results. But the system was not designed for different car speeds. Literature [12] established an EPAS mathematical model, and the simulation results showed that the strategy could achieve the desired characteristics, but the vehicle speed was not taken into account; the results had certain limitations [13, 14], using a sliding mode control that improved the system stability and anti-disturb capability but that increased the complexity of the control system, which set higher requirement of the computing power to the control ship. That is not beneficial to the promotion of products.

The aim of this study in EPAS is to control the electric motor to supply the appropriate assist torque to decrease the driver's steering effort in various speeds. The EPAS control must ensure the generation of the desired assist torque, a stable system with a large amount of assistance. The most important issue is electric motor tracking precisely the target current. To develop the electric motor current tracking performance, particle swarm optimization (PSO) algorithm is applied as tuning mechanism for fractional-order PID (FOPID) controller.

6. System modeling

The EPAS includes a torque sensor, which senses the action of the driver along with the action of the automobile; an ECU, which performs calculations on assisting force based on signals from the torque sensor; a motor, which creates turning power based on the output from ECU; and a reduction gear, which increases the turning force from the motor and transfers it to the steering system and pinion and rack (**Figure 8**).

The parameters associated with the rack model are M_r (steering tie rod mass), B_r (steering tie rod damping coefficient), R_s (radius of pinion steering), and K_r (tire spring rate) (**Table 5**).

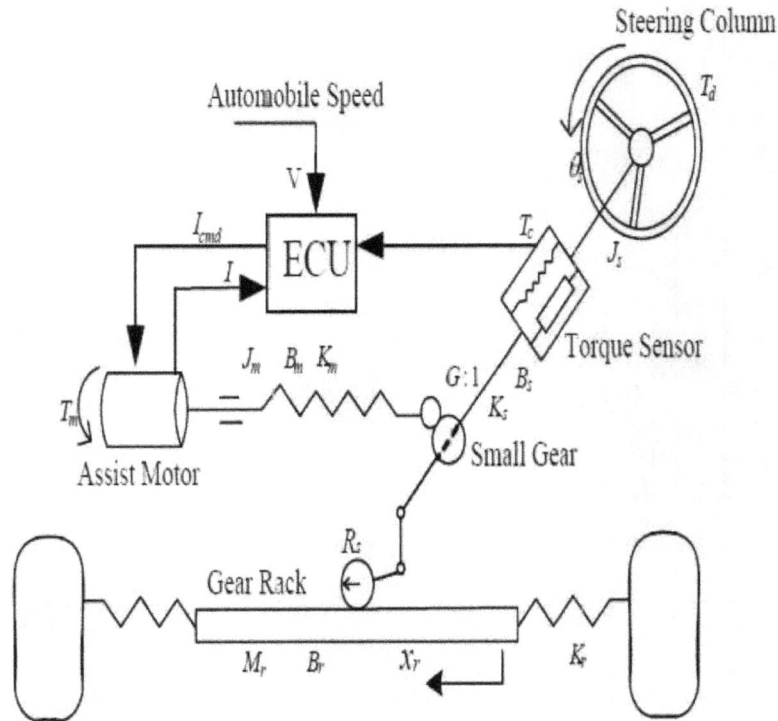

Figure 8. EAPS dynamic model.

Parameters	Symbols	Value	Units
Driving wheel moment of Inertia	J_s	0.04	$k_g.m^2$
Driving wheel damping	B_s	0.362	N.m.s.rad^{-1}
Pinion radius	R_s	0.0078	m
Rack and wheel assembly mass	M_r	32	K_g
Viscous rack damping	B_r	650.5	$N/\left(\frac{m}{s}\right)$
Motor gear ratio	G	16.5	
Motor stiffness	K_m	125	N.m.rad^{-1}
Motor inductance	L	0.0015	Henry
Motor resistance	R	0.15	Ohm
Motor torque constant	K_a	0.02	N.m.s.rad^{-1}
Motor EMF constant	K_b	0.02	v.s.rad^{-1}
Motor moment of Inertia	J_m	0.000452	$k_g.m^2$
Motor damping	B_m	0.003339	N.m.s.rad^{-1}
Steering column stiffness	K_s	115	N.m.s.rad^{-1}
Tire spring rate	K_r	91,061	N.m

Table 5. Parameters of EPAS system [14, 15].

7. EPAS controller

The function of ECU is to collect the torque sensor and the vehicle speed signal, select a suitable motor target current by an assist characteristic curve, execute a control by comparing with the feedback actual current, and then drive the DC motor.

8. Fractional-order PID (FOPID) controllers

Fractional-order PID (FOPID) controller denoted by $PI\lambda D\mu$ was proposed by Igor Podlubny [16] in 1997. It is an extension of conventional PID controller where λ and μ have fractional values. **Figure 9** shows the block diagram of FOPID controller. The fractional-order PID (FOPID) controller is a generalization of the PID controller. The transfer function of the controller is written by the equation below:

$$C(s) = K_p + \frac{K_i}{S^\lambda} + K_d S^\mu \tag{3}$$

where $K_p, K_i,$ and K_d are the proportional gain, integral gain, and derivative time constants, respectively, and λ and μ are fractional powers.

where μ and λ are an arbitrary real numbers. Taking $\mu = 1$ and $\lambda = 1$, a classical PID controller is obtained. Thus, FOPID controller generalizes the classical PID controller and expands it from point to plane as shown in **Figure 10**. This expansion provides us much more flexibility

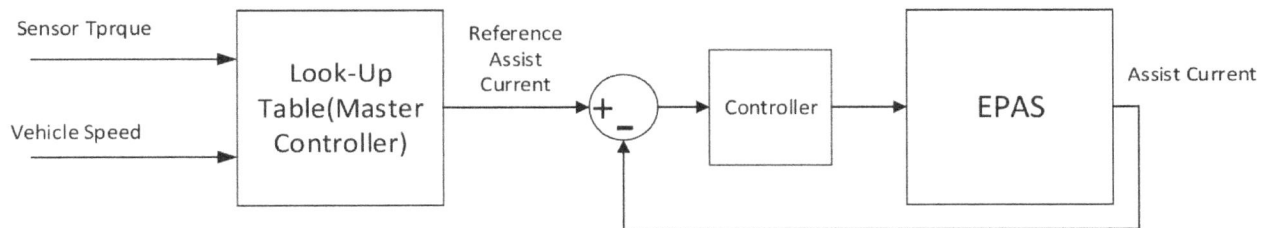

Figure 9. Fractional-order PID controller.

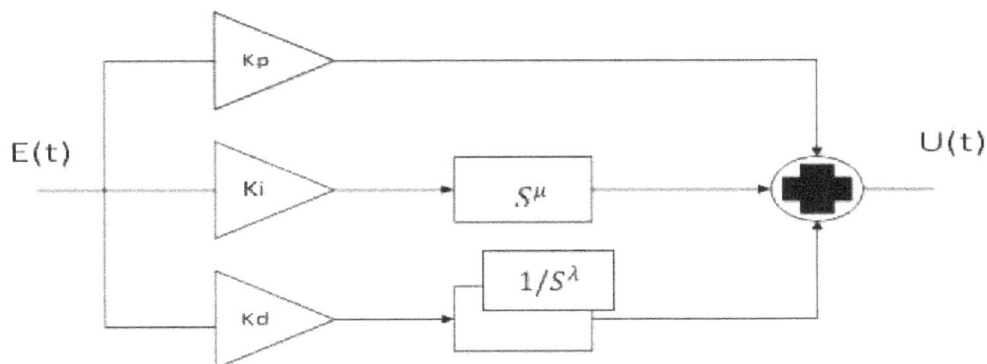

Figure 10. Control strategy.

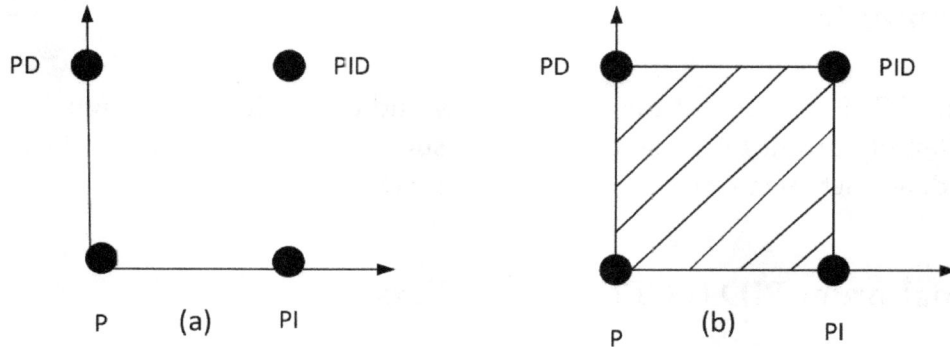

Figure 11. (a) Classical PID controller and (b) FOPID controller.

in designing PID controller and gives an opportunity to better adjust the dynamics of control system. This increases the robustness of the system and makes it more stable. However, with increase in parameters to be tuned, the optimization problem associated with the system becomes more difficult [17]. For achieving a certain performance, it is desired to develop a systematic algorithm for the FOPID optimization as shown in **Figure 11**.

9. Simulation results

Figure 12 shows an open-loop response of system, as depicted in the figure below, that motor current cannot follow the step unit. The close-loop unit step response of EPAS system using classical PID controller and PSO-FOPID are shown in **Figure 13**.

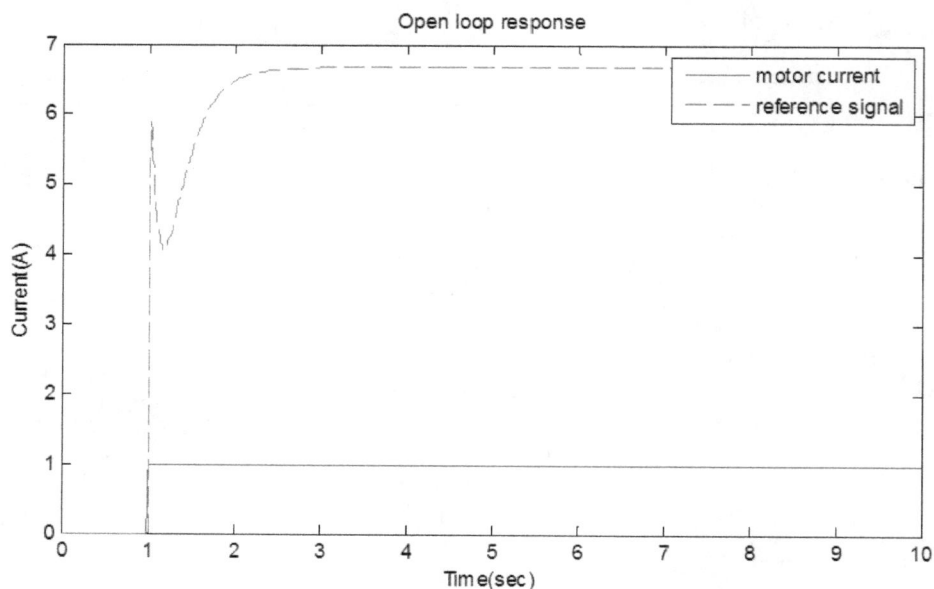

Figure 12. Unit step response of EPAS open-loop system.

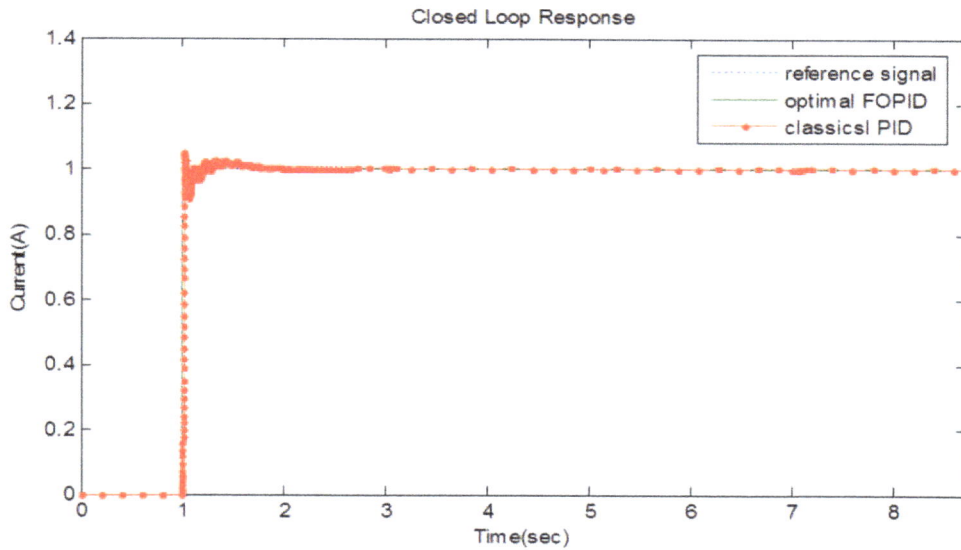

Figure 13. Unit step response of EPAS system using classical PID and optimal FOPID.

10. Controlled system in different speeds and different driver torques

Figure 14 shows three signals (i) motor current tracking of the look-up table, (ii) driving wheel input torque and (iii) the electric motor output current. The vehicle speed is 20km/h while the driver input torque is represented as a sine wave of 9 N.m amplitude.

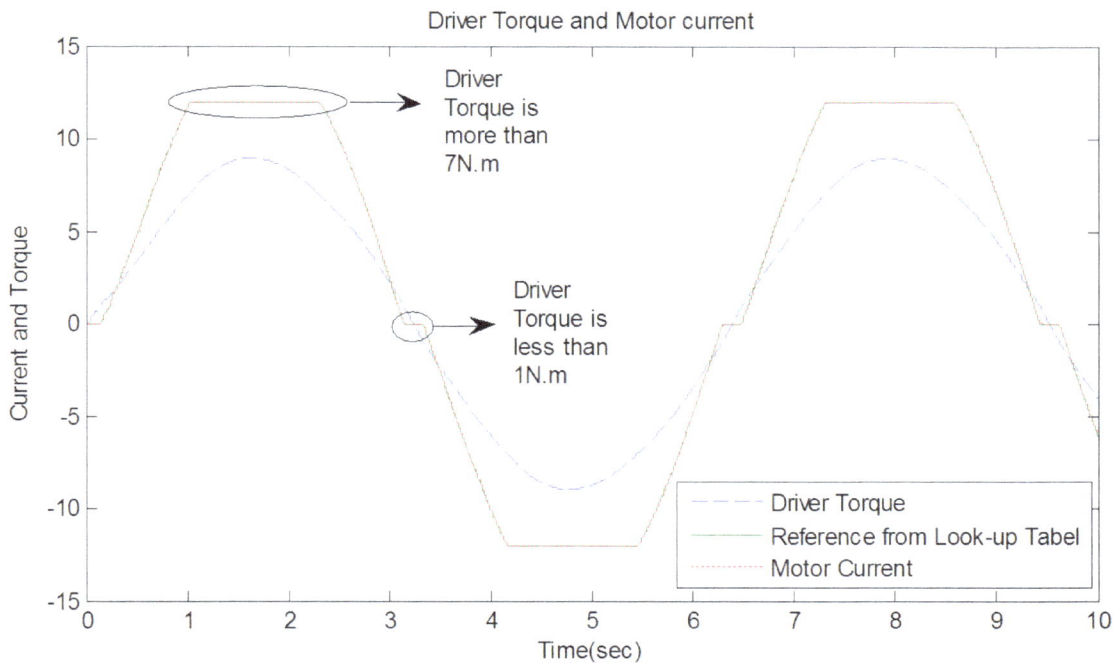

Figure 14. Driver torque and motor current in 20 km/h.

When the input torque of driving wheel T_d is less than the threshold value $T_{d0} = 1N.m$, motor does not provide power, so the assist current would be zero, when T_d is between $T_{d0} = 1N.m$ and $T_{dmax} = 7N.m$, motor current has a rising with T_d, and it depends on the car speed. When T_d is above T_{dmax}, the motor output is a constant torque.

11. Conclusion

In this study, a design of fuzzy-PID controller for BBW system is presented. In addition to that, a design structure for BBW is proposed which helps to elaborate a principle of work of the suggested BBW system. The braking mechanism and operation of BBW system are grasped and realized by obtaining mathematical derivation of the brake system based on quarter car model. Two controller algorithms based on PID and fuzzy-PID controllers are then implemented to check the validity of mathematical derivation on the one side and to operate braking mechanism of BBW on the other side. The simulation result which is conducted on different road types and conditions shows that fuzzy-PID controller is a superior and outstanding controller as compared to PID controller, where the fuzzy-PID controller assists to reduce stopping vehicle time 60% and the most important thing is the ability of fuzzy-PID controller to improve the system performance by eliminating steady-state error to zero. Besides, the result analysis and investigation demonstrate that larger adhesion characteristics lead to produce larger brake force which in turn assists to reduce vehicle stopping time.

For EPAS system, FOPID (fractional-order PID) controller has been presented, and it was tuned to control the motor current. All simulations for the whole EPAS system are implemented by MATLAB/Simulink software showing a comparison of classical PID and optimal PID tracking performance. PSO algorithm has been implemented to find optimal values of FOPID parameters. From the simulation results, it fulfills the control objectives and achieves good assistant in different speeds.

Author details

Ameer Sider[1]* and Mohd Khair Hassan[2,3]*

*Address all correspondence to: asider@birzeit.edu and khair@upm.edu.my

1 Department of Mechanical and Mechatronics Engineering, Faculty of Engineering, Birzeit University, Birzeit, Palestine

2 Department of Electrical and Electronic Engineering, Faculty of Engineering, Universiti Putra Malaysia, Selangor, Malaysia

3 Institute of Advanced Technology, Faculty of Engineering, Universiti Putra Malaysia, Selangor, Malaysia

References

[1] Mingfei Y. Brake-by-wire System design and simulation. In: International Conference of Computer Science and Electronics Engineering (ICCSEE), 2012. Hangzhou; 2012. pp. 248-251

[2] Xiang W, Richardson PC, Zhao C, Mohammad S. Automobile brake-by-wire control system design and analysis. IEEE Transactions on Vehicular Technology. 2008;**57**(1):138-145

[3] Saric S, Bab-Hadiashar A, Hoseinnezhad R. Clamp-force estimation for a brake-by-wire system: A sensor-fusion approach. IEEE Transactions on Vehicular Technology. 2008;**57**:778-786

[4] Lin WS, Tang TE. Active safety diagnosis of brake-by-wire systems with unscented Kalman filter. In: Proceedings of the 2010 IEEE International Conference on Vehicular Electronics and Safety (ICVES); QingDao, China; 2010. pp. 1-6

[5] Hoseinnezhad R, Bab-Hadiashar A. Recent patents on measurement and estimation in brake-by-wire technology. Recent Patents on Electrical and Electronic Engineering. 2009;**2**:54-64

[6] Burton AW. Innovation drivers for electric power-assisted steering. IEEE Control Systems. 2003;**23**(6):30-39

[7] Chen X. Optimal control of electric power-assisted steering system. [MSc Thesis], Windsor, Ontario, Canada; 2005

[8] Chun-Hua H. Modeling and simulation of automotive electric power steering system. Intelligent Information Technology Application. Shanghai China. IITA'08. 20-22 Dec 2008; 2008;**3**:436-439

[9] Shriwastava R, Diagavane M. Electric power steering with Permanent magnet synchronous motor drive used in automotive application. In: Proceedings of the 2011 1st International Conference on in Electrical Energy Systems (ICEES); Chennai India; 2011

[10] Chen X, Li K. Robust control of electric power-assisted steering system. In: Vehicle Power and Propulsion. Chicago, IL, USA: IEEE; 2005

[11] Qun Z, Juhua H. Modeling and simulation of the electric power steering system. In: Circuits, Communications and Systems. 2009. PACCS'09. Pacific-Asia Conference on. Chengdu, China: IEEE; 2009

[12] Parmar M, Hung JY. A sensorless optimal control system for an automotive electric power assist steering system. IEEE Transactions on Industrial Electronics. 2004;**51**(2):290-298

[13] Marouf A, et al. Control of electric power assisted steering system using sliding mode control. In: Intelligent Transportation Systems (ITSC), 2011 14th International IEEE Conference on. Nagoya, Japan· 2011

[14] Marouf A et al. A new control strategy of an electric-power-assisted steering system. IEEE Transactions on Vehicular Technology. 2012;**61**(8):3574-3589

[15] Hassan M et al. Optimal design of electric power assisted steering system (EPAS) using GA-PID method. Procedia Engineering. 2012;**41**:614-621

[16] Dzielinski A, Sierociuk D. Simulation and experimental tools for fractional order control education. IFAC Proceedings Volumes. 2008;**41**(2):11654-11659

[17] Padhee S et al. A novel evolutionary tuning method for fractional order PID controller. International Journal of Soft Computing and Engineering (IJSCE). 2011;**1**(3):1-9

Permissions

The contributors of this book come from diverse backgrounds, making this book a truly international effort. This book will bring forth new frontiers with its revolutionizing research information and detailed analysis of the nascent developments around the world.

We would like to thank all the contributing authors for lending their expertise to make the book truly unique. They have played a crucial role in the development of this book. Without their invaluable contributions this book wouldn't have been possible. They have made vital efforts to compile up to date information on the varied aspects of this subject to make this book a valuable addition to the collection of many professionals and students.

This book was conceptualized with the vision of imparting up-to-date information and advanced data in this field. To ensure the same, a matchless editorial board was set up. Every individual on the board went through rigorous rounds of assessment to prove their worth. After which they invested a large part of their time researching and compiling the most relevant data for our readers.

The editorial board has been involved in producing this book since its inception. They have spent rigorous hours researching and exploring the diverse topics which have resulted in the successful publishing of this book. They have passed on their knowledge of decades through this book. To expedite this challenging task, the publisher supported the team at every step. A small team of assistant editors was also appointed to further simplify the editing procedure and attain best results for the readers.

Apart from the editorial board, the designing team has also invested a significant amount of their time in understanding the subject and creating the most relevant covers. They scrutinized every image to scout for the most suitable representation of the subject and create an appropriate cover for the book.

The publishing team has been an ardent support to the editorial, designing and production team. Their endless efforts to recruit the best for this project, has resulted in the accomplishment of this book. They are a veteran in the field of academics and their pool of knowledge is as vast as their experience in printing. Their expertise and guidance has proved useful at every step. Their uncompromising quality standards have made this book an exceptional effort. Their encouragement from time to time has been an inspiration for everyone.

The publisher and the editorial board hope that this book will prove to be a valuable piece of knowledge for researchers, students, practitioners and scholars across the globe.

List of Contributors

Roxana-Elena Tudoroiu and Sorin-Mihai Radu
University of Petrosani, Petrosani, Romania

Mohammed Zaheeruddin
University Concordia from Montreal, Montreal, Canada

Nicolae Tudoroiu
John Abbott College, Saint-Anne-de-Bellevue, Canada

Zeki Omaç
Electrical and Electronics Engineering, University of Munzur, Tunceli, Turkey

Mehmet Polat
Mechatronics Engineering, University of Fırat, Elazığ, Turkey

Mustafa Kaya
Digital Forensics Engineering, University of Fırat, Elazığ, Turkey

Eyyüp Öksüztepe and Haluk Eren
School of Aviation, University of Fırat, Elazığ, Turkey

Merve Yıldırım and Hasan Kürüm
Electrical and Electronics Engineering, University of Fırat, Elazığ, Turkey

Sorin Ioan Deaconu, Marcel Topor and Lucian Nicolae Tutelea
Politehnica University Timisoara, Romania

Vasile Horga
Technical University, Iasi, Romania

Fabrizio Marignetti
University of Cassino and Southern Lazio, Italy

Ilie Nuca
Technical University of Moldova, Chisinau, Republic of Moldova

Michela Longo and Federica Foiadelli
Department of Energy, Politecnico di Milano, Milan, Italy

Wahiba Yaïci
CanmetENERGY Research Centre, Natural Resources Canada, Ottawa, Ontario, Canada

Ananchai Ukaew
Development and Research of Innovative Vehicle Engineering (DRIVE) Center, Faculty of Engineering, Naresuan University, Thailand

Eiman Elbanhawy
Oxford Brookes University, Oxford, UK

Shinji Kajiwara
Faculty of Science and Engineering, Kindai University, Higashioska, Osaka, Japan

Carlos López-Torres, Antoni Garcia-Espinosa and Jordi-Roger Riba
Electrical Engineering, Universitat Politècnica de Catalunya (UPC), Terrassa, Spain

Yi-Hsien Chiang and Wu-Yang Sean
Department of Environmental Engineering, Chung Yuan Christian University, Taiwan

Ameer Sider
Department of Mechanical and Mechatronics Engineering, Faculty of Engineering, Birzeit University, Birzeit, Palestine

Mohd Khair Hassan
Department of Electrical and Electronic Engineering, Faculty of Engineering, Universiti Putra Malaysia, Selangor, Malaysia
Institute of Advanced Technology, Faculty of Engineering, Universiti Putra Malaysia, Selangor, Malaysia

Index

www.ingramcontent.com/pod-product-compliance
Lightning Source LLC
Chambersburg PA
CBHW080633200326

41458CB00013B/4609